Antiterrorism and Homeland Defense

ACS SYMPOSIUM SERIES **980**

Antiterrorism and Homeland Defense

Polymers and Materials

John G. Reynolds, Editor
Lawrence Livermore National Laboratory

Glenn E. Lawson, Editor
Naval Surface Warfare Center

Carolyn J. Koester, Editor
Lawrence Livermore National Laboratory

Sponsored by the
Division of Polymeric Materials: Science and Engineering, Inc.

American Chemical Society, Washington, DC

ISBN: 978-0-8412-3964-7

The paper used in this publication meets the minimum requirements of American National Standard for Information Sciences—Permanence of Paper for Printed Library Materials, ANSI Z39.48-1984.

Copyright © 2007 American Chemical Society

Distributed by Oxford University Press

All Rights Reserved. Reprographic copying beyond that permitted by Sections 107 or 108 of the U.S. Copyright Act is allowed for internal use only, provided that a per-chapter fee of $36.50 plus $0.75 per page is paid to the Copyright Clearance Center, Inc., 222 Rosewood Drive, Danvers, MA 01923, USA. Republication or reproduction for sale of pages in this book is permitted only under license from ACS. Direct these and other permission requests to ACS Copyright Office, Publications Division, 1155 16th Street, N.W., Washington, DC 20036.

The citation of trade names and/or names of manufacturers in this publication is not to be construed as an endorsement or as approval by ACS of the commercial products or services referenced herein; nor should the mere reference herein to any drawing, specification, chemical process, or other data be regarded as a license or as a conveyance of any right or permission to the holder, reader, or any other person or corporation, to manufacture, reproduce, use, or sell any patented invention or copyrighted work that may in any way be related thereto. Registered names, trademarks, etc., used in this publication, even without specific indication thereof, are not to be considered unprotected by law.

PRINTED IN THE UNITED STATES OF AMERICA

Foreword

The ACS Symposium Series was first published in 1974 to provide a mechanism for publishing symposia quickly in book form. The purpose of the series is to publish timely, comprehensive books developed from ACS sponsored symposia based on current scientific research. Occasionally, books are developed from symposia sponsored by other organizations when the topic is of keen interest to the chemistry audience.

Before agreeing to publish a book, the proposed table of contents is reviewed for appropriate and comprehensive coverage and for interest to the audience. Some papers may be excluded to better focus the book; others may be added to provide comprehensiveness. When appropriate, overview or introductory chapters are added. Drafts of chapters are peer-reviewed prior to final acceptance or rejection, and manuscripts are prepared in camera-ready format.

As a rule, only original research papers and original review papers are included in the volumes. Verbatim reproductions of previously published papers are not accepted.

ACS Books Department

Contents

Introduction

1. Polymers and Materials for Antiterrorism and Homeland Defense: An Overview..................3
   John G. Reynolds and Glenn E. Lawson

Chemical Detection

2. Synthesis and Spectroscopic Characterization of Molecularly Imprinted Polymer Phosphonate Sensors..................19
   G. E. Southard, K. A. Van Houten, Edward W. Ott Jr., and G. M. Murray

3. Development of an Enzyme-Based Photoluminescent Porous Silicon Detector for Chemical Warfare Agents..................39
   Bradley R. Hart, Sonia E. Létant, Staci R. Kane, Masood Z. Hadi, Sharon J. Shields, Tu-Chen Cheng, Vipin K. Rastogi, J. Del Eckels, and John G. Reynolds

4. Optical Enzyme-Based Sensors for Reagentless Detection of Chemical Analytes..................57
   Brandy Johnson-White and H. James Harmon

5. Design of Sorbent Hydrogen Bond Acidic Polycarbosilanes for Chemical Sensor Applications..................71
   Eric J. Houser, Duane L. Simonson, Jennifer L. Stepnowski, Mike R. Papantonakis, Stuart K. Ross, Stanley V. Stepnowski, Eric S. Snow, Keith F. Perkins, Chet Bryant, Peter LaPuma, Gary Hook, and R. Andrew McGill

6. Non-Aqueous Polymer Gels with Broad Temperature
   Performance..89
   Joseph L. Lenhart, Phillip J. Cole, Burcu Unal, and
   Ronald C. Hedden

7. Detection of Toxic Chemicals for Homeland Security Using
   Polyaniline Nanofibers...101
   Shabnam Virji, Richard B. Kaner, and Bruce H. Weiller

8. Applications of Nanoparticles in Scintillation Detectors....................117
   Suree S. Brown, Adam J. Rondinone, and Sheng Dai

Biological Detection

9. A Comparison of Insulator-Based Dielectrophoretic Devices for
   the Monitoring and Separation of Waterborne Pathogens as a
   Function of Microfabrication Technique..133
   Gregory J. McGraw, Michael Kanouff, Joseph T. Ceremuga,
   Rafael V. Davalos, Blanca H. Lapizco-Encinas, Petra Mela,
   Renee Shediac, John D. Brazzle, John T. Hachman,
   Gregory J. Fiechtner, Eric B. Cummings, Yolanda Fintschenko,
   and Blake A. Simmons

10. Design and Synthesis of Dendritic Tethers for the
    Immobilization of Antibodies for the Detection of
    Class A Bioterror Pathogens..159
    Charles W. Spangler, Brenda D. Spangler, E. Scott Tarter,
    and Zhiyong Suo

Decontamination and Protection

11. Amphiphilic Polymers with Potent Antibacterial Activity..................175
    M. Firat Ilker, Gregory N. Tew, and E. Bryan Coughlin

12. Catalysts for Aerobic Decontamination of Chemical Warfare
    Agents under Ambient Conditions..198
    Craig L. Hill, Nelya M. Okun, Daniel A. Hillesheim,
    and Yurii V. Geletii

13. Ultrastable Nanocapsules from Headgroup Polymerizable
    Divinylbenzamide Phosphoethanolamine...210
    Glenn E. Lawson and Alok Singh

14. Nanoencapsulation of Organophosphorus Acid Anhydrolase with Mesoporous Materials for Chemical Agent Decontamination in Organic Solvents..233
Kate K. Ong, Tu-Chen Cheng, Ray Yin, Hua Dong, Jian-Min Yuan, and Yen Wei

Indexes

Author Index...255

Subject Index...257

Introduction

Chapter 1

Polymers and Materials for Antiterrorism and Homeland Defense: An Overview

John G. Reynolds[1] and Glenn E. Lawson[2]

[1]Forensic Science Center, Lawrence Livermore National Laboratory,
P.O. Box 808, L–178, Livermore, CA 94551 (reynolds3@llnl.gov)
[2]Chemical/Biological Defense Center, 17320 Dahlgren Road, Building 1480,
Code Z21, Naval Surface Warfare Division, Dahlgren, VA 22448–5000
(glenn.lawson@navy.mil)

Polymers and materials play key roles in national security through detection and decontamination addressing chemical, biological, radiological, nuclear and explosives threats. Proposed detection and decontamination methods utilize polymers and other materials to combat terrorist threats. Because today's detectors are not sufficiently sensitive and selective and decontamination agents are not selective enough for all threats and every scenario, research is being conducted to bridge the scientific and technical gaps. Collected in this volume are papers that elucidate specific efforts in developing new polymers and new materials that can be used as platforms in detectors, as the matrix to incorporate specific detection sites, as recognition elements for detection, as detectors themselves, as decontamination agents and as key components in detection systems. The chapters are divided into the categories of chemical detection, biological detection, and decontamination. Although these groupings are primarily based on the applications, much of the design of the polymers and materials can be broadened into other pertinent detection and decontamination scenarios.

A variety of polymeric materials and the methodology to produce them are described. These polymers include cross-

© 2008 American Chemical Society

linked divinyl benzene-substituted methacrylate polymers, polycarbosilanes, non-aqueous chemically cross-linked polybutadiene gels, conducting polyaniline nanofibers, organically doped polystyrene and polyvinyltoluene, electroplated polymer cast resins, amphiphilic functionalized norbornene polymers, cross-linked divinyl-benzamide phospholipids, silica and organo-silyl polymers. Other materials include metal chelating complexes, functionalized porous silicon, siloxyl immobilized enzymes with porphyrins, polycarbosilanes, quantum dots and nano-crystalline oxides, dendritic complexing sites, amphiphilic functionalized norbornene polymers, reactive glass surfaces, and self assembled monolayers.

Introduction

Since, the tragedy that occurred on September 11, 2001, there have been dramatic increases in research activities to develop chemical and biological detection systems for anti-terrorism and homeland defense. To protect military personnel and the general public, chemical and biological warfare detection systems are rapidly becoming an essential part of the homeland security and defense strategies of the United States. Critical research is being conducted in government, academic, and industrial laboratories. The majority of this research activity has been in the collection, detection and mitigation of chemical, biological, nuclear and explosive (CBNE) materials related to weapons of mass destruction (WMD). The aim of this research is concentrated in two different directions; one is concerned with protection and decontamination on the battlefield, the other, protection within civilian populated centers. Battlefield protection and decontamination encompasses rapid detection, neutralization and removal of chemical and biological agents from military vehicles, equipment, personnel, and facilities (1–3). Protection of civilian centers encompasses rapid detection and removal of chemical and biological agents from public buildings, equipment, and civilians, to concentrations that are lower than those required for military applications. At this time, new detection systems are necessary because current technology is not sufficient in warning of the presence of such weapons. New mitigation methods are necessary for protecting military personnel and the civilian population from attack, as well as securing facilities from contamination.

The chemical warfare agents (Figure 1) of main focus in this book are HD, VX, GB, and GD. HD is a blistering agent that attacks the mucous membranes and is lethal at high doses. VX, GB, and GD are nerve agents that have the

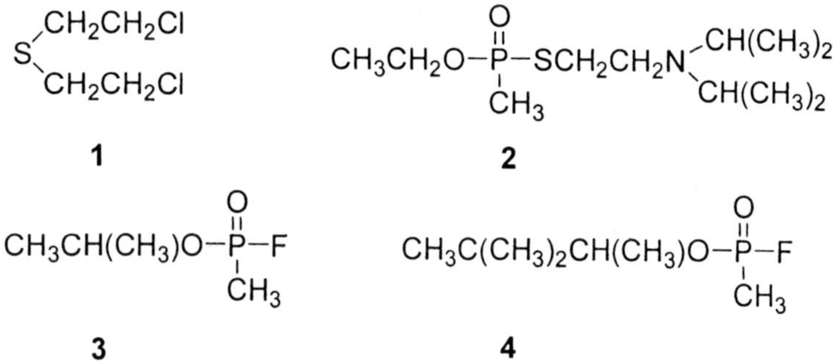

Figure 1. Chemical structures of warfare agents sulfur mustard (1), VX (2), GB (3), and GD (4).

ability to stop respiratory and nerve functions, as well as kill in only minutes, even in fairly low concentrations (4, 5).

Due to the toxic nature of the chemical warfare agents, compounds used during initial investigations of new materials and sensors are often simulants or surrogates. Simulants mimic one or more of the physical properties (for example, vapor pressure) of a chemical warfare agent and have significantly lesser toxicities than those of the chemical agents. Surrogates are compounds that are similar in chemical structure (i.e., contain common functional groups) to chemical agents and, for this reason, might have significant, albeit lesser, toxicities. For example organophosphorus pesticides are often used as surrogates for the organophosphorus nerve agents. The use of simulants and surrogates makes it possible to more safely study decontamination mechanisms, like oxidation, at pentavalent phosphorus atoms as substitutes for nerve agents. The results of these studies can then be extrapolated to chemical warfare agents. Although, simulants and surrogates can facilitate detoxification and decontamination studies, critical final testing must be performed with actual agents. Such testing is extremely dangerous, time consuming and expensive.

Detection and decontamination of biological threat agents is also covered in this book. Biological agents cover a broad range of pathogens that either attack organisms directly or produce large toxins, which also debilitate and kill. Detection and decontamination is often studied using surrogates, as in the case of chemical warfare agents. Table I, shows a partial list of these threats from the Centers for Disease Control and Prevention (6).

In order for a chemical or biological detection system to be successful, it must have the ability to specifically and selectively identify a preferred target immediately. An ideal decontamination system should neutralize the target within minutes with a high efficacy, be non-toxic to the user, be non-corrosive

Table I. Partial List of Threats from the Select Agent Program of the Centers for Disease Control and Prevention

Select Agents	Toxins
Bacillus anthracis	Ricin
Brucella	Saxotoxin
Francisella tularensis	Tetrodotoxin
Marburg Virus	Botulinum toxins
Yersinia pestis	

to surfaces and the surrounding environment, and be simple to apply by either the military personnel or civilians.

To achieve these goals, one avenue of great promise is through development of new materials. Ongoing, applied research of new materials has led to a broad range of collectors and specific chemical detectors that can be coupled to devices. For example, development of new materials has led to collectors that can be coupled to devices such as surface acoustic wave (SAW) sensors and ion mobility spectrometers (IMS) to produce specific chemical detectors for nerve agents and explosives. Development of new materials has also shown the way to new breathable, fabrics for protective equipment that is resistant to chemical attack for first responders. Enzyme-based decontamination can be facilitated by biomolecules or polymeric vesicles. Hence, development of new materials and novel strategies can be exploited for the decontamination of sensitive items ranging from multi-million dollar military aircraft and ship-based electronics and optics to everyday computers and art.

This book has been divided into three areas: chemical detection, biological detection, and decontamination. The subject matter in the chapters include cross-linked divinyl benzene-substituted methacrylate polymers (Chapter 2), porous silicon (Chapter 3), reactive glass surfaces (Chapter 4), polycarbosilanes (Chapter 5), non-aqueous, chemically cross-linked polybutadiene gels (Chapter 6), conducting polyaniline nanofibers (Chapter 7), organically doped polystyrene and polyvinyltoluene (Chapter 8), electroplated polymer cast resins (Chapter 9), self assembled monolayers (Chapter 10), amphiphilic functionalized norbornene polymers (Chapter 11), transition metal substituted polyoxometalates (POMs) (Chapter 12), cross-linked divinyl-benzamide phospholipids (Chapter 13), and silica and organo silyl polymers (Chapter 14).

The purpose of this symposium book is to provide an overview of the subject of detection and decontamination from a materials viewpoint. Chapters focus on esoteric research and development on key new materials. The development of these materials is discussed on both the fundamental and applied levels. This chapter is only a brief overview of the material contributed to this symposium book. The reader is referred to individual chapters and references contained within for a comprehensive discussion on the subject matter.

Chemical Detection

For a sensor to detect a target molecule and be usable in a detection system, it must have the following features: a recognition mechanism, such as a binding site that is specific to the target molecule, free access of the target to the this site, a method that can indicate when the target is interacting, and, preferably, an escape route for the target molecule when detection is complete, so that the detection system can be reused. There are many different means for implementing these features and innovations in any of these areas can greatly improve the selectivity and/or sensitivity of a detection system.

Thus, much research has gone into designing detection systems by a variety of different approaches. Some concentrate on the binding site and optimize selectivity. Developing the target recognition site often requires some type of manipulation or processing of a material to have different and specialized properties. Materials may be functionalized to attach to or add into surface materials to make a specific binding or recognition site. Sometimes this requires complex functionalization chemistry or chelation chemistry. In other cases, the material itself is manipulated to form the recognition site, such as molecular imprinting in polymeric matrices, or to perform the recognition, such as in insulator-based dielectrophoresis. Many materials have been modified to be recognition sites or serve as recognition sites through manipulation of physical properties.

Other approaches to chemical detection systems concentrate on the transduction of the binding event, highlighting the sensitivity. In this case, the function of the material is to report the interaction of the sensor with the target. Photoluminescence, for example, is a desirable property because changes in the emitted light can be manipulated by materials development to indicate the interaction of the target with the recognition site.

Regardless of what approach is taken towards the development or performance optimization of a detection system, the selection of appropriate materials will impact every part of sensor design and fabrication. The challenge to researchers is to design systems for chemical and biological detection that are extremely selective to the target and give a very strong indication to extremely small quantities of the target molecule. The following paragraphs describe the approaches taken by different researchers and the significance of their work.

Synthesis and Spectroscopic Characterization of Molecularly Imprinted Polymer Phosphonate Sensors, Chapter 2

For molecular imprinting of polymers, the binding site can be constructed by polymerizing the polymer around a target molecule. This target molecule is then removed after polymerization is complete. Ideally, the cavity left will be an exact image of the target molecule, giving selectivity towards the target molecule upon re-exposure. However, in practice, interfering molecules that are similar in chemical structure to the target analyte will also interact with the polymer and more sophisticated methods are necessary to provide the selectivity needed for detection.

An innovative approach taken by Southard and coworkers is to incorporate a Eu(III) binding site into the molecularly imprinted polymer. This binding site offers some specificity for organophosphorus compounds, such as nerve agents. The incorporation of specific lanthanide metal ligand complexes into polymers, through conventional free radical polymerization or Reversible Addition Fragmentation Transfer (RAFT) polymerization, aids in the detection of the organophosphorus compounds. Eu(III) ligand complexes will show different optical properties with and without binding of the target molecule. Thus, the immobilized system will exhibit enough spectroscopic difference upon binding of the target that it is easily detected by measuring luminescence changes.

Development of an Enzyme-Based Photoluminescent Porous Silicon Detector for Chemical Warfare Agents, Chapter 3

Another method of developing an analyte-specific sensor is by covalent attachment of an enzyme onto a surface. Enzymes can be very specific towards detection of target molecules. Detection comes through monitoring the formation of the enzymatic product, through inhibition of, or possibly through changes in the enzyme structure during enzymatic activity. This specificity can exceed that of the antibody-antigen interactions. One drawback of using enzymes is finding the appropriate enzyme with activity towards the specific target to be detected. Another drawback is that enzymes are expensive to produce; however, new advances are overcoming target sensitivity and production issues and are making enzyme use much more viable.

Porous silicon is a desirable surface material for sensors because of its photoluminescence properties. With excitation, the porous silicon surface can luminesce at a variety of easily detected wavelengths. Through various methods, such as quenching, the surface can show some evidence of interaction with a target molecule. However, plain porous silicon lacks the specificity needed to detect target molecules, and is subject to non-specific interferences.

To enhance the specificity towards target molecules, specific sites for interaction have been developed. A select group of enzymes show selectivity towards organophosphorus compounds, including nerve agents, through the action of hydrolysis. These enzymes present potential for detection if they can be immobilized on the surface of the porous silicon and the enzymatic activity affects the optical properties of the porous silicon. Hart and coworkers have used hydrosilation to form a robust link between an organophosphorus anhydrolase enzyme and a porous silicon surface, without seriously affecting the optical properties. They show that this sensor successfully detects *p*-nitrophenyl-soman.

Optical Enzyme-Based Sensors for Reagentless Detection of Chemical Analytes, Chapter 4

Johnson-White and Harmon study a novel detection technique using enzymes and highly-colored porphyrin compounds, in concert, for sensing of organophosphorus compounds. Immobilized enzymes are treated with porphyrin compounds that interact at the binding site of the enzyme. Porphyrins are intensely colored compounds with very strong extinction coefficients, ideal for use in optical systems. When exposed to the target, the enzyme releases the porphyrin in favor of the target. The decrease in the concentration of the bound porphyrin is measured through the UV-vis spectrum and quantification is achieved by comparing this value to target-free surface. This system could detect GB at concentrations below those that are considered to be immediately dangerous to life and health (IDLH).

Design of Sorbent Hydrogen Bond Acidic Polycarbosilanes for Chemical Sensor Applications, Chapter 5

Because of their ability to collect basic vapors, such as organophosphorus compounds, hydrogen-bond acidic polymers are useful sorbents for detector platforms. However, these polymers are not commercially available. Houser and coworkers describe recent advances in the preparation of hydrogen-bond acidic polycarbosilanes to enhance their sensitivity towards organophosphorus compounds. Polycarbosilane polymers were prepared from various substituted monomers with Pt-catalyzed or thermal ring opening reactions. These parent polymers were then reacted with different fluoro-substituted compounds to give a variety of fluoro-substituted polymers. Fluoroalcohol-substituted polycarbosilanes provided a higher sensitivity toward hydrogen-bond basic vapors than the related polysiloxane analogues; this was attributed to the absence of the hydrogen bond basic oxygen sites in the polycarbosilane backbone. These materials were characterized and tested in a variety of sensor applications (surface acoustic wave (SAW), single wall carbon nanotube networks (CNN) sensors, and a commercial attenuated total reflectance Fourier Transform infrared (ATR/FTIR) system).

Non-Aqueous Polymer Gels with Broad Temperature Performance, Chapter 6

Polymer gels offer potential for sensing because their properties can be tuned by varying the polymer type, solvent type, and solvent loading. In addition, small molecule additives and fillers can be incorporated into the gel formulation to further enhance their properties. Hydrogels belong to a limited class of polymer gels that show great utility to biological applications due to near-room-temperature responsiveness. While hydrogels are useful for the biological and medical community, and the thermo-responsiveness of these gels can be exploited for certain devices, their use is limited for applications such as homeland security due to limited operating temperature range because of the moderate freezing point and high volatility of water in the gel. For this reason, a need exists for polymer gels that maintain properties and performance over a broad range of temperatures.

Lenhart and coworkers describe the design, synthesis and characterization of non-aqueous, polybutadiene gels that exhibit the polymer gel criteria listed above with specific performance over a temperature range of -70 to 70 °C. These gels are stable and the methods described can be extended to other systems, which might eventually lead to other potential applications including homeland security detection systems, micro-devices, electronics encapsulation, energy storage devices, information protection, protective coatings, and controlled lubrication layers.

Detection of Toxic Chemicals for Homeland Security Using Polyaniline Nanofibers, Chapter 7

As discussed previously, many chemical threats have acid/base properties, Thus, polyaniline films, which have been demonstrated as sensor materials for acid/base compounds, are logical substrates for use in chemical detection systems. In addition, polyaniline films have the electrical properties of a conducting polymer, so that they can be used for transduction, as well as molecular recognition. However, due to sensitivity, time response, and weakness, these sensors have limitations. For these reasons, Virji and coauthors have worked to improve the polyaniline films and incorporate them into much more robust systems.

Polyaniline nanofibers are synthesized through a simple, non-templated method, based on a modification of the classic chemical oxidative polymerization of aniline that makes nanofibers of uniform diameters. These materials respond significantly better than conventional films. To make the polyaniline even better for sensor materials, the nanofibers are dispensed into water and are combined with metals salts into a composite material. These materials showed superior performance for the detection of several toxic gases (HCl, NH_3, N_2H_4, and H_2S) as compared to conventional polyaniline materials.

Application of Nanoparticles in Scintillation Detectors, Chapter 8

Advanced radiation detectors, especially for neutron and gamma radiation, are important for detecting "dirty" bombs, monitoring U-235 and Pu-239 for nuclear arms control/verification, uncovering nuclear smuggling, special nuclear material (SNM) detection, and waste characterization, as well as for fundamental research in nuclear physics, solid-state physics, chemistry, biology, and neutron radiography. The scintillation process, occurring when the energy of ionizing radiation is absorbed by certain crystalline inorganic or organic materials and resulting in the emission of UV-vis light from the absorbing materials, remains as one of the most useful methods for detecting ionizing radiation. The key to the development of advanced radiation detectors lies in the synthesis and characterization of efficient scintillation materials. Polymers can be utilized in the design of innovative radiation detectors.

The ideal scintillation material should meet the following criteria: high scintillation efficiency, wide range of linear energy conversion, transparency to the wavelength of its own emission, short emission decay time, good optical quality, index of refraction near that of glass, and emission spectrum matched to the spectral sensitivity of the detector, and robust durability. Commercially available inorganic scintillation materials possess many of the desired properties. However, their particle sizes, in the range of microns, lead to the opacity of scintillation detectors and, consequently, lowered scintillation efficiencies.

During the past two decades, there have been extensive investigations on the use of inorganic nanocrystals in scintillation detectors. Brown and coworkers present a survey of representative organic nanoparticles and inorganic nanocrystals embedded in various matrices, along with their performances in beta, alpha, or neutron detection. This study leads towards the ability to identify the most suitable nanoparticle-based scintillation materials for specific radiation detection purposes, and to a better understanding of how certain nanoparticle-based scintillators behave under radiation interactions.

Biological Detection

A Comparison of Insulator-Based Dielectrophoretic Devices for the Monitoring and Separation of Water-Borne Pathogens as a Function of Microfabrication Technique, Chapter 9

The potential of micro-fluidic devices for separation and detection of several types of target compounds has been shown over the years. Lab-on-a chip devices have been fabricated with a variety of separation capabilities coupled with detection modalities. Polymer-based micro-fluidic devices have an added advantage because they can be fabricated by mass production techniques. For example, injection molding and hot embossing, which have long histories of commercial fabrication applications, can be used.

Polymeric, micro-fluidic devices have been utilized by McGraw and coworkers to separate and concentrate biological materials. Using insulator-based dielectrophoretic devices, water-borne bacteria, spores, and inert materials were separated, allowing for detection and removal. In the formation of these devices, masters from two different base materials, glass and silicon, were used. The glass master was formed using chemical isotropic etching and the silicon master was formed using glass reactive ion anisotropic etching. Each master imparted different properties in the fabrication process. The reactive ion etching produced relatively straight sidewalls, whereas the chemical etching produced tapered sidewalls. These masters were sputter coated and then Ni was electroplated on the surface. The microfluidic devices were fabricated using injection molding with the commercially available Zeonor® 1060R resin. The devices exhibited similar structure derived from the parent master, and the devices differed in similar ways to how their masters differed.

The performance of these polymer micro-fluidic devices was examined as a function of their structural properties and the target morphology. The polymeric devices exhibit essentially the same behavior and performance in insulator-based dielectrophoresis as their glass counterparts. This indicates the potential for mass production of these devices to address the needs of high-throughput processes.

Design and Synthesis of Dendritic Tethers for the Immobilization of Antibodies for the Detection of Class A Bioterror Pathogens, Chapter 10

The antibody-complement antigen relationship is one of the more specific interactions for recognizing target materials. To optimally use this interaction for detection, the antibody must be immobilized on a surface. This immobilization has to be performed in such a way as to leave the antibody intact, as well as not to inhibit access. A number of ways have been attempted to do this immobilization. A novel technique has been developed by Spangler and coworkers that employs monodispersed, dendritic, heterobifunctional tether capture agents that can be derivatized for attachment as self-assembled monolayers on a variety of surfaces, such as gold. These prepared surfaces can then be used in surface plasmon resonance and quartz crystal microbalance instruments. The synthetic procedures needed to assemble such systems are described. The utility of this approach is demonstrated by showing the behavior of anti-BSA immunoglobulin (antibody) binding to surface-immobilized bovine serum albumin (BSA). This work is expected to lead to the production of better immobilization agents.

Decontamination and Protection

Amphiphilic Polymers with Potent Antibacterial Activity, Chapter 11

Decontamination is an important part of countering WMD. New materials can play a role if decontamination activity can be designed into the materials. Ilker and coworkers describe the synthesis and activities of new amphiphilic polymers with potent antibacterial activity. These polymers are synthesized with domains that have polar and non-polar character. The character and size of each domain can be tuned independently and locked into the repeating units. Living ring opening metathesis polymerization (ROMP) of these monomers is used for polymerization. This approach allows synthetic control of the structure of the resulting materials. Polymers with a large range of molecular weights with narrow distributions can be synthesized. The start of the monomer design is based on the widely used norbornene derivatives. Both hydrophobic and hydrophilic groups can be incorporated into the nonbornene structure, allowing for the construction of amphiphilic properties. These simple polymers represent a new approach to the development of nontoxic, broad-spectrum antimicrobials and have significant potential for applications in bioterrorism defense.

Amphiphilic polynorbornene derivatives were tested for lipid membrane disruption activities using liposomes. Water-soluble, amphiphilic, cationic polynorbornenes derivatives exhibited the highest activities, showing the best promise for anti-bacterial activity. These were further tested for their antibacterial activities in growth inhibition assays and hemolytic activities against human red-blood cells. This allowed for the determination of selectivity of the polymers for bacterial over mammalian cells. It was found that highly selective polymers with non-hemolytic, anti-bacterial activities could be designed by tuning the overall hydrophobicity of the polymer.

Catalysts for Aerobic Decontamination of Chemical Warfare Agents Under Ambient Conditions, Chapter 12

Destruction and decontamination of chemical warfare agents is important in many scenarios, including facility restoration and personnel protection. There are many approaches to decontamination, with the route dependent on the agent as well as the application. Hydrolysis is an effective decontamination method for fluorophosphonate agents such as sarin (see Figure 1). Oxidation, often using peroxides such as H_2O_2, is another decontamination method and is applicable to agents such as sulfur mustard and VX. An innovative, and perhaps a less severe, oxidative approach is to use catalysis with air as the oxidation source.

Hill and coworkers have developed oxidation catalysts, $Fe^{III}[H(ONO_2)_2]PW_{11}O_{39}^{5-} \cdot HNO_3$ and $Ag_2[(CH_3C(CH_2O)_3)_2V_6O_{13}$, to attack agents, such as mustard and VX, using oxygen. For the destruction of sulfur mustards, these catalysts are designed to convert the thioether S to sulfoxide and stop. This provides an advantage over other commonly-used oxidative methods, which completely oxidize the sulfur mustard to its sulfone, a vesicant that is as potent as sulfur mustard. Several catalysts have been developed, but, the most active to date is $Fe^{III}[H(ONO_2)_2]PW_{11}O_{39}^{5-} \cdot HNO_3$, an iron-substituted polyoxometalate. The activity of this catalyst is compared to a variety of other catalysts for the oxidation of 2-chloroethyl ethyl sulfide, a surrogate for sulfur mustards.

Ultrastable Nanocapsules from Headgroup Polymerizable Divinylbenzamide Phosphoethanolamine, Chapter 13

Synthetic vesicles have potential application for drug delivery and nanomaterial release. The three-dimensional structure of the vesicle is ideal for encapsulation of materials that could be useful for decontamination and detection in support of homeland security. In addition, the ability to modify and vary the chemical composition of the surfaces of the vesicle membranes gives vesicle encapsulation the potential for a wide variety of applications. However, systematic formation of highly stable vesicles, necessary for broadening applications and commercial production, remains elusive. Vesicles tend to be thermodynamically unstable, can readily undergo secondary processes like fusion and disintegration, and are generally destroyed by many organic solvents. All these issues limit vesicle use.

Lawson and Singh report on efforts to overcome these weaknesses in vesicle structure. Employing techniques to cross-link head groups, much more stable vesicles are formed, as witnessed by their survival of freeze-drying and re-suspension, and resistance to decomposition in both protic and aprotic organic solvents. The flexibility of these materials is also documented by the ability to release and incorporate hydrophilic and hydrophobic materials without structure disruption. The synthesis process is fast and can be performed at room temperature, making it ideal for a variety of applications.

Nanoencapsulation of Organophosphorus Acid Anhydrolase (OPAA) with Mesoporous Materials for Chemical Agent Decontamination in Organic Solvents, Chapter 14

In addition to being used in chemical sensors, as described in Chapter 3, immobilized organophosphorus hydrolase enzymes can also be used to deactivate or decontaminate nerve agents. Silica is an excellent choice for immobilization because the enzyme can be condensed into the structure through the sol-gel process, producing a very stable solid material. In addition, the bulk properties of the silica can be modified through precursor materials. And, post processing assists in facilitating access of the target molecule to the material, as well as manipulating the bulk properties of that material.

One issue in the sol-gel process is forming materials of controlled porosity, necessary for access of the target molecules to the enzymatic site. Ong and coworkers have used a non-surfactant, templated approach to synthesize materials of reasonably high porosity with the enzyme present. In addition, the silica properties were also varied using silica and organo-siloxane precursors for the sol-gel. The immobilization of OPAA in mesoporous materials significantly increased the stability of OPAA against denaturing in the presence of organic solvents. The application of these materials to decontamination was measured through the hydrolysis of diisopropyl fluorophosphates.

Concluding Remarks

Due to the relative ease of modifying their physicochemical properties, polymers continue to play a major role the detection arena. Research will continue to increase their specificity, or ability to selectively bind to target analytes, and sensitivity, their ability to transduce interactions with target analytes. Successful research in these areas will eventually increase the range of applications to which these materials can be applied. In the future, polymers will also play an important role in decontamination and protection from various threat agents. Materials for decontamination and protection will be designed to neutralize the target instantaneously, with a high efficacy, be non-corrosive to surfaces and the surrounding environment, be simple to apply by either the military personnel or civilians, and ultimately be inexpensive to produce.

Acknowledgment

This work was performed under the auspices of the U.S. Department of Energy by University of California Lawrence Livermore National Laboratory under contract No. W-7405-Eng-48.

References

1. Yang, Y.-C.; Baker, J. A; Ward, R. *Chem. Rev.* **1992**, 1729-1743.
2. Thompson, J. H.; Schwartz, M. *Evaluation of the Resistance of Standard Air Force Paint to Liquid Toxic Agent Sorption*, ARCSL-TM-79016, 1979.

3. Thompson, J. H.; Day, S. E.; Schwartz, M.; Keck, C. H. *Development of a Chemical Agent Resistant Coating*; ARCSL-TM-80017, 1980.
4. Papirmeister, B.; Feister, A. J.; Robinson, S. I.; Ford, R. D. *Medical Defense Against Mustard Gas: Toxic Mechanisms and Pharmacological Implications;* CRC Press, Inc.: Boca Raton, FL, 1991.
5. Benschop, H. P.; de Jong, L. P. A. *Acc. Chem. Res.* **1988**, *21*, 368-374.
6. http://www.cdc.gov/od/sap/; Department of Health and Human Services, Centers for Disease Control and Prevention, Select Agent Program website, updated October 27, 2005.

Chemical Detection

Chapter 2

Synthesis and Spectroscopic Characterization of Molecularly Imprinted Polymer Phosphonate Sensors

G. E. Southard, K. A. Van Houten, Edward W. Ott Jr., and G. M. Murray[*]

Applied Physics Lab, The Johns Hopkins University, Laurel, MD 20723

> Molecularly imprinted polymers (MIPs) that are capable of sensing specific organophosphorus compounds, such as pinacolyl methylphosphonate (PMP), by luminescence have been synthesized and characterized. The polymers have been synthesized using conventional free radical polymerization and using Reversible Addition Fragmentation Transfer (RAFT) polymerization. The RAFT polymers exhibited many advantages over conventional free radical processes but are more difficult to make porous.

The preparation of molecularly imprinted polymers (MIPs) normally employs conventional free radical polymerization. While this is a relatively simple process, it is prone to problems that can affect the imprinting site. RAFT polymerization can overcome some of the problems associated with conventional free radical polymerizations and offers an opportunity to make block copolymers, thus localizing the crosslinking of the polymer around the imprinting site.

The Synthesis of Imprinted Polymers for Luminescence Detection

Molecular imprinting is a process for making selective binding sites in synthetic polymers (*1*). The process may be approached by designing the recognition site or by simply choosing monomers that may have favorable interactions with the imprinting molecule. To successfully apply the methodology to chemical sensing requires the designed approach. The process involves building a complex of an imprint molecule and complementary polymerizable ligands. At least one of the molecular complements must exhibit a discernable physical change associated with binding. This change in property can be any measurable quantity, but a change in luminescence is the most sensitive and selective analytical technique. By copolymerizing the complexes with a matrix monomer and a suitable level of crosslinking monomer, the imprint complex becomes bound in a polymeric network. The network must be mechanically and chemically processed to liberate the imprinting species and create the binding site (*2-4*). The design of the binding site requires chemical insight. These insights are derived from studies of molecular recognition and self-assembly and include considerations of molecular geometry, size and shape, as well as molecule-to-ligand thermodynamic affinity.

For the purpose of sensing, a reporting molecule must be a part of the cavity to indicate when the target molecule (e.g., pinacolyl methylphosphonate) is bound. The best reporters generate fluorescence, since photoluminescence is highly sensitive and adds greater selectivity through the two wavelengths of light that must be employed. Compounds like pinacolyl methylphosphonate are complex organic molecules that exhibit a variety of chemical functional groups as substituents. Many of these groups possess an affinity for metal ions in solution and can form complexes. By using these natural affinities, metal incorporating organic copolymers can be made that have a high thermodynamic affinity for binding molecules. The organic matrix of the polymer improves vapor detection since it can selectively condense organic vapor from air, as is the case in solid phase micro-extraction (SPME) methods. The judicious choice of a metal ion with useful spectroscopic properties, in addition to a high thermodynamic binding affinity, results in the formation of highly selective and sensitive sensors.

The mechanism to detect pinacolyl methylphosphonate is to discern its effect of the luminescence of a lanthanide ion, Eu^{3+} (*5*). The optical absorption and emission spectra of the triply charged free lanthanide ions, which are assignable to f→f transitions, generally consist of very narrow lines (0.1–0.01 nm). Organic compounds with affinity to form metal ion complexes are called ligands. When ligands are placed around the lanthanide ions, the spectral lines are shifted in position (wavelength), they may be split (usually ~1.0 nm), and they are often altered in relative intensity (some of them may even be eliminated) due to electrostatic interactions called the Stark Effect. The exact character and degree of these changes are specific to each different ligand environment, the major factors being the coordination number, the coordination symmetry (geometry), and the type(s) of ligands. Even though the ligand field is sufficient to produce these shifts, splits, and intensity alterations, the ligand to f electron coupling is generally not adequate to result in broadening of the lines. Another consequence of the weak coupling and the inter-configurational nature of f→f transitions, radiative lifetimes of lanthanide ions in compounds are quite long, on the order of milliseconds, simplifying time resolved measurements that enhance sensitivity. By a judicious choice of coordinating ligand, the absorptivity or quantum efficiency can be enhanced and a broad band source, such as a light emitting diode (LED) can be used. Of the lanthanide ions, the Eu^{3+} ion has the simplest set of splitting possibilities and most favorable level structure for sensitization. These spectroscopic properties make Eu^{3+} ions attractive for chemical sensing (6).

Thus, the exchange of one ligand for another results in spectral changes that are useful for sensing. In the case of a vapor sensor, the analyte need only displace a solvent molecule (typically water) to bind to Eu^{3+} ion. This process is facilitated by the imprinted cavity geometry and the affinity of organic polymers to absorb organic molecules. In order to develop a binding site with appropriate characteristics for sensing, several candidate complexes must be screened. The selection process is based on finding a complex that has a spectral shift with complexation and that will absorb light from an inexpensive and low power source. The complexing species also has to bind Eu^{3+} tightly enough to keep it from being removed from the polymer when the imprinting species is removed. In order to satisfy these criteria, we selected β-diketones. Once the pair of complexes with and without imprinting ion were shown to exhibit the correct spectroscopic properties, it was necessary to functionalize the complexing ligands with vinyl groups. However, it is vital that the ligands be polymerizable, so that the imprinted site will be bound to the polymer.

Incorporation of the functionalized complex into the polymer matrix required some experimentation (7). The metal ion complexes may have limited solubility in the monomer solution. This can be addressed by adding an appropriate solvent to the mixture. The solvent has the further purpose of acting to help increase the porosity of the finished polymer. This simplifies the removal of the imprint molecule and increases the speed of the sensor response. Typically, a range of complex loading, crosslinking, and solvent addition are examined for optimal optical response.

Experimental Details

Reagents

4-Bromobenzoylacetone was synthesized by a modified Shea method (8). The new bromine substituted β-diketones were synthesized by modified literature procedures (9–11). The 1-acetyl-4-bromonaphthalene was synthesized by standard Friedel-Crafts acylation of 1-bromonaphthalene with acetyl chloride and aluminum chloride. The synthesis of the vinyl substituted β-diketones, see Figure 1, has been reported elsewhere (19). All chemicals were provided by Sigma-Aldrich unless otherwise stated, and were used without further purification.

Instrumentation

An in-house detection system was employed for screening lanthanide complexes that includes: an Ar Ion Laser, Model 543 Head and Model 170 Power Supply, (Omnichrome, Chino, CA) and an $f/4$, 0.5 meter monochromator (Chromex, Albuquerque, NM) equipped with a Model ST-6 CCD detector (Santa Barbara Instrument Group, Inc., Santa Barbara, CA). KestrelSpec software (Rhea Corp. Wilmington, DE) was used to operate the CCD and record the compound luminescence. A Cary 50 UV-vis spectrophotometer (Varian, Walnut Creek, CA) was used to obtain absorbance spectra. The luminescence titrations were obtained using a Model QM-2 Fluorimeter/Phosphorimeter, (Photon Technologies International, Monmouth, NJ). Thermogravimetry was performed using a Model SDT 2960 Simultaneous DSC-TGA (TA Instruments, New Castle, DE). A Hewlett-Packard Model 5400 ICP-MS (Yokogawa Analytical Systems, Tokyo, Japan) was used to verify metal concentrations in all sample solutions. NMR was performed using either a Bruker AC-200 MHz spectrometer or a Model EFT 90 MHz spectrometer (Anasazi Instruments, Indianapolis, IN). The purity of synthesized organics was established using a Model QP 5050A GC/MS (Shimadzu, Columbia, MD). The Heck coupling reaction was performed with either a Model HC 677 100-mL reactor (Parr Instruments, Moline, IL) or an LC series 300-mL reactor (Pressure Products Industries, Warminster, PA).

Synthesis of HDBNTFA (8): Dithiobenzoic acid 1-(4-(4,4,4-trifluoro-butane-1,3-dione)-naphthalen-1-yl)-ethyl ester (HDBNTFA)

Compound **6a** (2.92 g, 10 mmol), dithiobenzoic acid (1.54 g, 10 mmol), and carbon tetrachloride (6 mL) were placed together into a 15-mL, round-bottomed flask equipped with a reflux condenser under an argon atmosphere. The reaction was heated to 70 °C for 16 h when a 2nd aliquot of dithiobenzoic acid (0.77 g, 5 mmol) and the reaction was continued for another 4 h. The solvent was removed by vacuum and the final product was isolated by column chromatography through silica gel with 60/40 hexanes/chloroform as eluent to give a viscous red oil (2.2 g, 50% yield). ^{1}H-NMR (90 MHz, 25 °C, CDCl$_3$): δ 8.52–8.45 (dd, 2 H), 8.09–7.36 (bm, 9 H), 6.80 (s, 1 H), 6.11–6.04 (q, 1 H), 2.00–1.92 (d, 3 H).

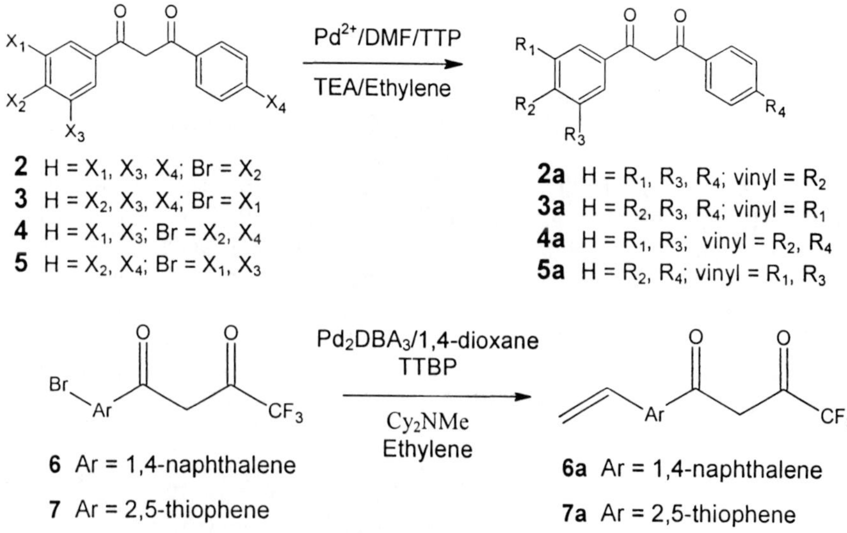

Figure 1. Steps involved in the synthesis of vinyl-β-diketones.

Synthesis of (DBNTFA)$_3$Eu·3H$_2$O (9)

Compound **8** (1.0 g, 2.24 mmol) was dissolved in THF (5 mL) in a 15-mL, round-bottomed flask and 1.0 M sodium hydroxide (2.46 mL) was added, dropwise. A solution of europium chloride hexahydrate (0.274 g, 0.75 mmol) in water (2 mL) was added and the flask was equipped with a reflux condenser. The reaction was heated to reflux for 3 h before excess methanol was added to end the reaction. The precipitate was removed by filtration, dried, dissolved in ether, filtered again, and precipitated into hexanes. The precipitate was isolated by filtration, which gave a red solid (750 mg, 66% yield). Anal. (found/calc for C$_{69}$H$_{54}$EuF$_9$O$_9$S$_6$): C 53.55 (53.73), H 3.56 (3.53).

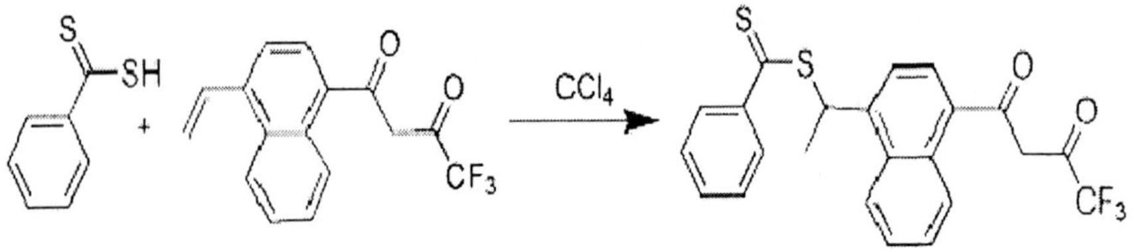

Figure 2. Preparation of the dithioester ligand (8).

Polymer Synthesis

General procedure for RAFT bulk polymerizations of 9 with methacrylics

Ethylene glycol dimethacrylate (EGDMA, 16 mmol), methyl methacrylate (MMA, 8 mmol), toluene (4 mL), Wako V-65 (0.044 mmol), pinacolyl methylphosphonate (0.029 mmol), and **9** (0.029 mmol) were placed into a disposable glass reaction flask equipped with a stir bar. The solution was subjected to three freeze/pump/thaw cycles with argon backfill. The solution was placed into an oil bath heated to 60 °C for 18 h before the solvent and unreacted monomer were removed by heating to 60 °C while under vacuum (0.5 torr) for 4 h. The salmon-colored polymer was ground with a freezer mill to a fine powder.

General procedure for RAFT bulk polymerizations of 9 with styrenics

Divinylbenzene (DVB, 22 mmol), styrene (11 mmol), toluene (4 mL), Wako V-65 (0.044 mmol), PMP (0.029 mmol), and **9** (0.029 mmol) were placed into a disposable glass reaction flask equipped with a stir bar. The solution was subjected to three freeze/pump/thaw cycles with argon backfill. The solution was placed into an oil bath heated to 60 °C for 18 h before the solvent and unreacted monomer were removed by heating to 60 °C while under vacuum (0.5 torr) for 4 h. The salmon-colored polymer was ground with a freezer mill to a fine powder.

There were difficulties in getting good complexation and adduct formation with vinyl substituted β-diketone ligands as discussed below. A path to an imprinting complex that does not involve ordinary free radical polymerization was sought. RAFT polymerization is a free radical process that is controlled and ensures the formation of high polymer. The process involves a chain transfer moiety usually based on a dithioester. Because this functionality does not involve conjugation to the β-diketone aromatic ring, it does not interfere with the ligands complexing ability, Figure 2. Ligands prepared with the dithioester substituents seen in Figure 3 formed stable complexes with europium(III) and formed stable adduct complexes with PMP.

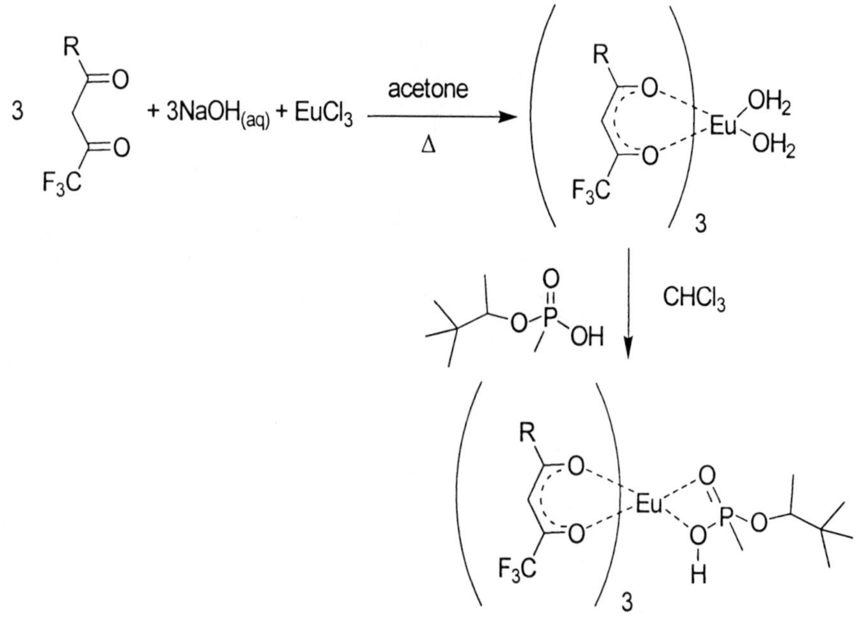

Figure 3. Synthesis of the tris (β-diketone) europium(III) complex, followed by a description of PMP adduct formation in chloroform.

β-Diketone Complexes

β-Diketone complexes of lanthanides have a long history of use due to their stability and utility as optical sensitizers (*12, 13*). Both properties are useful in sensor applications. When the project began, the choice of excitation sources was limited to LEDs operating at blue wavelengths of 470 nm and higher. In order to sensitize a Eu^{3+} complex to such a long wavelength, a ligand with extensive conjugation would be needed. Thus, the initial choice for a complexing ligand was dibenzoylmethane. The addition of one or two vinyl substituents would increase the conjugation and gave us the best chance of using blue LED excitation. The alternate choice was benzoylacetone, a ligand with less conjugation but with a less sterically-hindered route to benzoate addition. Interestingly, the substitution of a vinyl group on the aromatic rings of the β-diketone ligands reduced their ability to form complexes and require different methodologies. The first and simplest methodology was to prepare a mixed ligand complex with two fluorinated ligands and the one vinyl-substituted β-diketone that still complexes europium(III), 3-vinyldibenzoylmethane. In this manner, the fluorinated ligands made europium(III) hard enough to form phosphonate adducts and the 3-vinyldibenzoylmethane made the complex polymerizable. The presence of water of hydration was verified using thermogravimetric analysis. The thenoyltrifluoroacetone complex had two water molecules of hydration, as expected, to give europium(III) the normal coordination number of nine. The vinyl-naphthoyltrifluoroacetone complex showed an additional two molecules of water per complex that are suspected to be lattice water.

As stated above, the dithioester-substituted ligand behaved in the same manner as the unsubstituted-trifluoromethyl ligands. The formation of the europium complex and the suspected route of adduct formation is shown in Figures 4 and 5.

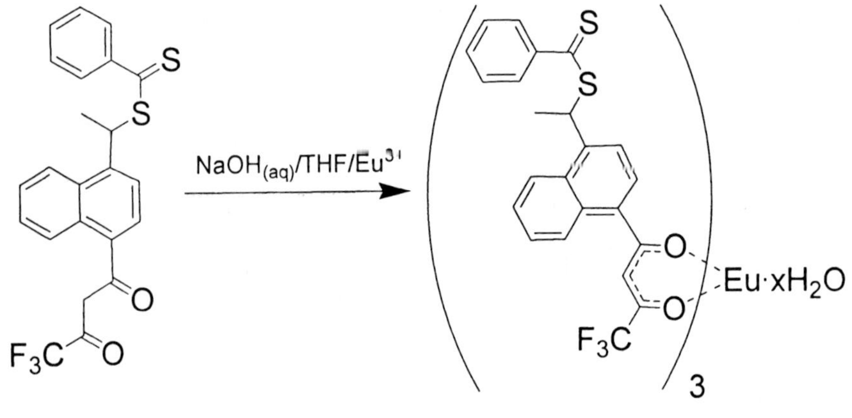

Figure 4. Preparation of the tris chelate.

MIPs were prepared for PMP, using a mixed ligand approach (3-vinyldibenzoylmethane and naphthoyltrifluoroacetone) in a matrix of MMA and EGDMA, using toluene as a porogen. The amount of the crosslinker, EGDMA, was 80 percent by weight. The polymers produced were ground in a cryo-grinder to a fine powder. Several cleaning methods were characterized so as to best remove the imprint molecule without compromising the integrity of the MIP. The washes were evaluated using gas chromatography/mass spectrometry. Of the systems tested, it was determined that unadulterated methanol was the best solvent for removing PMP without compromising polymer integrity.

The mixed ligand approach was followed by the RAFT approach. RAFT polymerization is a free radical process that is controlled and ensures the

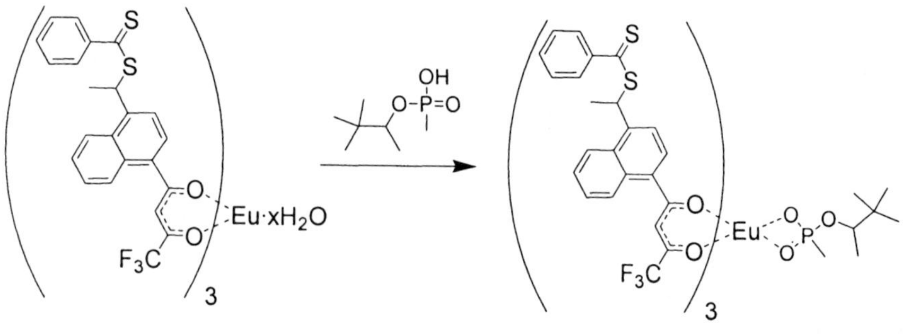

Figure 5. Conversion to the imprint complex.

formation of high polymer with a narrow molecular weight distribution (*14–18*). The process involves a chain transfer agent usually based on a dithioester. The process provides for a highly controlled system allowing the use of a broad spectrum of monomers and a highly controlled topology and morphology. The polymers were made using both styrene (Sty) and MMA as matrix monomers. The porosity of the resultant polymers was limited. The procedures were repeated using methoxyethanol in place of toluene as the solvent. The polymers prepared with methoxyethanol ground easier and were more miscible with xylene. The polymerization procedure for the styrenic RAFT polymer is illustrated in Figure 6 (*20*).

Characterization of β-diketone Complexes

Complexes of Eu^{3+} with β-diketone coordinating ligands were routinely investigated using laser excited luminescence spectra obtained with a high resolution spectrograph. The intent was to see if spectral changes occurred to the $^7F_2 \leftarrow {}^5D_0$ (hypersensitive) luminescence band when the imprint molecule was included in the complex. Because this band exhibits the greatest sensitivity to changes in shape (coordination geometry) and intensity (charge density), it is the band most useful for creating a sensor.

Based on the screening of ligands described above, the first goal in ligand synthesis was to produce 1,3-diphenyl-2-(4-vinylbenzyl)-1,3-propanedione or, descriptively, a dibenzoylmethane substituted on each end with a vinyl group. The rational was that this ligand would absorb blue light from an LED source and be the best choice for imprinting since each end would be linked to the polymer. When the Eu^{3+} complex was synthesized, the spectrum did show absorbance of light around 430 nm, Figure 7. While this was shy of the goal of 470 nm, it was synonymous with the commercial availability of 430 nm LEDs. These results were promising and polymers were made using this system. It was soon apparent that having the vinyl groups on either end of the ligand had changed the complexation constant and made the complex susceptible to hydrolysis.

At this point, it was discovered that new LEDs were commercially available operating at wavelengths shorter than 430 nm. These near-UV light sources had advantages for the sensor in terms of increasing the Stokes shift for luminescence and simplifying the choice of ligand monomers. We reasoned that a less sterically hindered coordinator, such as benzoylacetone, where the second benzene ring is replaced by a smaller methyl group, might be more desirable and show more significant spectral changes. Thus, the next coordinator tested was benzoylacetone.

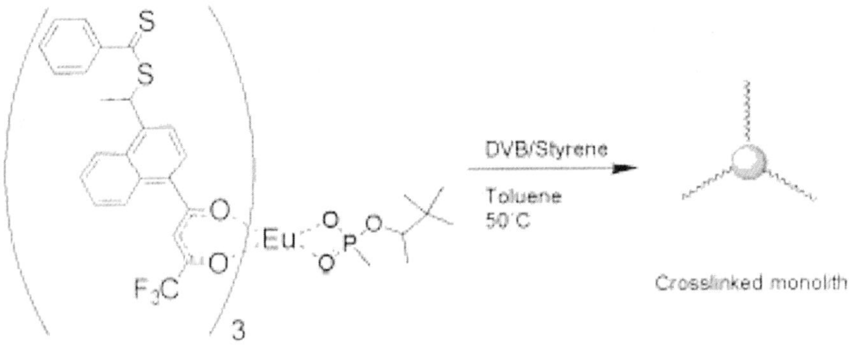

Figure 6. RAFT polymerization of the MIP for PMP. (Reprinted from reference 20, with permission from Elsevier.)

Results and Discussion

Metal ion complexes can form an imprint site that exhibits both a spectral signature and can be expected to survive incorporation into a polymer. A solution analogue of the imprint site can be prepared and the utility for use in imprinting may be established. For a d-block complex, adduction may be characterized by obtaining absorbance, or in some cases, luminescence spectra. The spectral changes that accompany adduct formation are either changes in band position or intensity. Lanthanide-based transduction complexes are usually based on luminescence intensity. If, as in the case of a lanthanide-based reporter, luminescence varies with the degree of analyte inclusion, a luminescence titration will reveal stoichiometry.

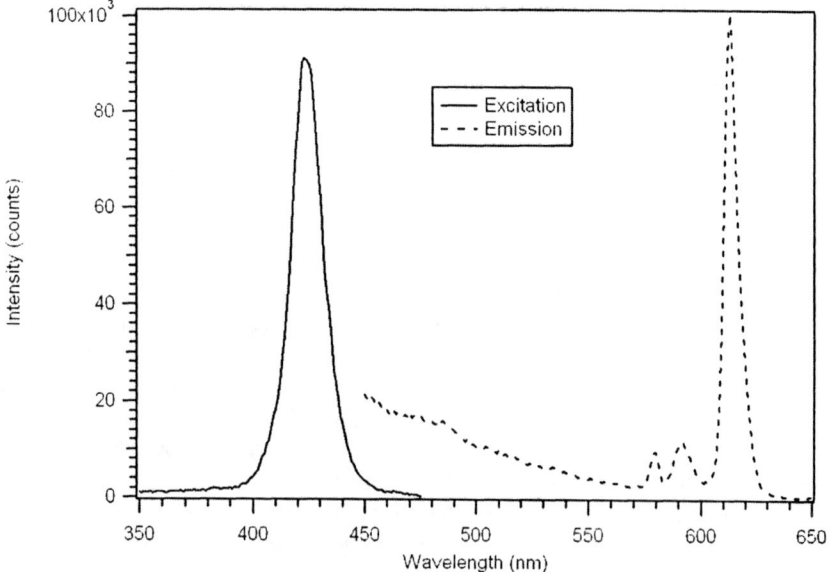

Figure 7. Luminescence spectra of tris (divinyldibenzoylmethanato) europium(III). (Reprinted from reference 20, with permission from Elsevier.)

We improved the Eu(III) binding site for organophosphates by replacing vinyl benzoate with vinyl β-diketones (2). A considerable amount of time and effort were devoted to synthesizing a variety of vinyl-substituted β-diketones (19); however, in the course of this work, it was discovered that some of the vinyl-substituted β-diketones exhibit significantly different complexing abilities relative to their unsubstituted relatives. The vinyl-substituted β-diketones with the vinyl groups in the 4 position or with multiple vinyl moieties were found to complex Eu^{3+} poorly, if at all, while the other vinyl-substituted β-diketones made complexes with Eu^{3+} only with some difficulty. Additionally, PMP would displace the non-fluorinated β-diketones ligand from the tris (β-diketone) europium(III) complexes instead of water, resulting in a decrease in luminescence. Fortunately, it was found that two β-diketones, thenoyltrifluoroacetone and naphthoyltrifluoroacetone, were sufficiently stable for the synthesis of organophosphate adducts. However, to produce polymers with thermodynamically and chemically stable complexes using these ligands, a mixed ligand approach was required. Hence, the PMP adducts and subsequent polymers were synthesized with 2:1 ratio of naphthoyltrifluoroacetone and 3-vinyldibenzoylmethane.

Although, a thioester-substituted β-diketone ligand was eventually chosen for the sensor, many vinyl-substituted β-diketones were found to be unsuitable for making stable europium complexes. For example, methylene substituted β-diketones exist as three tautomers, as observed by NMR spectroscopy. The most stable tautomer gives isolated π system, resulting in unusual chemistry. The ligands no longer complex well and do not sensitize luminescence. The resulting compounds were insensitive towards the analyte. The initial choice of ligand monomer was a vinyl-substituted form of dibenzoylmethane. Problems associated with this choice began when it was discovered it was not possible to couple vinyl-acetophenone to vinyl-methylbenzoate using sodium hydride. After the appropriate synthetic route was established, the potentially most attractive variant for imprinting, 4,4'-divinyldibenzoylmethane proved too sterically hindered to form viable complexes. It had been hoped that two connections to the polymer chain would provide the most stable binding site. The excitation spectrum of the europium(III) complex showed a maximum at 430 nm, a wavelength easily obtainable using a light emitting diode. When other forms of dibenzoylmethane were employed, the complexes partially dehydrated in organic solvents. While this was reversible, it was unexpected and caused confusion when attempting to interpret spectra.

Harry Brittain (21) developed a luminescence titration method for the determination of the stoichiometry of β-diketone europium(III) complexes and phosphate esters. In his work, adducts were formed in a 1:1 mole ratio for tris (β-diketone) europium(III) complexes having at least one trifluoromethyl substituent. We applied this method to determine the stoichiometry of the adduct of PMP and tris(naphthoyltrifluoroacetone) europium(III), $Eu(NTFA)_3$, in chloroform. The excitation wavelength was 360 nm and the slits of the excitation and emission monochromators were set at 2 nm. A triangular cell, instead of the 180° geometry used by Brittain, was used to reproduce the experiment in a commercial fluorimeter. So, instead of 3.0 mL aliquots, 1.5 mL aliquots were employed. The spectra produced by the addition of partial equivalents of PMP to $Eu(NTFA)_3$ up to 1.05 equivalents is presented in Figure 8. Unlike the phosphate esters studied by Brittain, addition of PMP beyond one equivalent causes a loss of β-diketone, as seen in Figure 9. Thus, instead of reaching a sustained maximum intensity of luminescence, the luminescence increases up to one equivalent and then decreases. The plot does show that the 1:1 complex has the greatest luminescence intensity. The complex should be suitable for an imprinted sensor, since, once formed, the amounts of PMP that the sensor will encounter should never approach the stoichiometric value.

The inclusion of a luminescent chromophore into an organic polymer may be complicated by background luminescence. When $Eu(NTFA)_3$ was incorporated into a Sty copolymer, the continuous wave (CW) luminescence spectrum showed a large background, Figure 10. This background is observed when using a broad band light source for excitation and not observed when the luminescence is excited by an Ar ion laser at 465.8 nm. Since our intent is to make small portable sensors, the ability to pump a broad, allowed ligand band with a small light source such as a LED, as opposed to pumping a sharp, weak europium(III) absorbance with a large laser, is desired. The background was

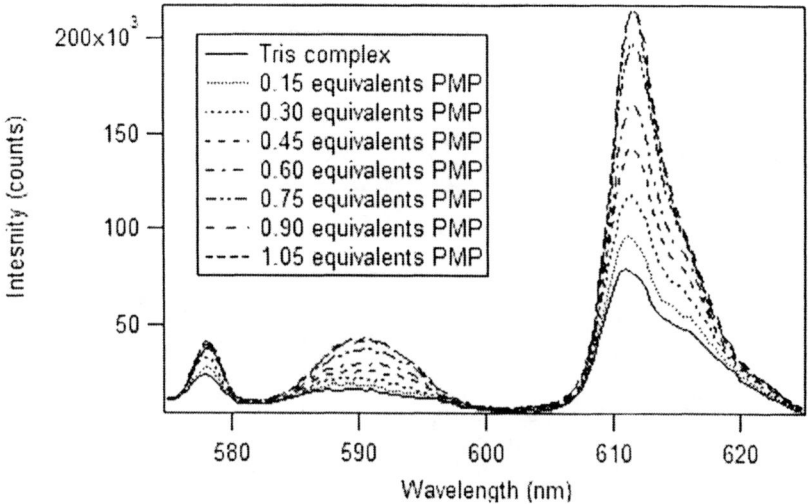

Figure 8. Luminescence spectra of Eu(NTFA)$_3$ with addition of PMP in chloroform. (Reprinted from reference 20, with permission from Elsevier.)

eliminated by using a pulsed light source and gated detection. The excitation spectra are not very different except there are some spikes due to the pulsed lamp's output. The emission spectrum now has a flat background, showing the absence of scattered light and matrix fluorescence. The time-resolved spectra were obtained by delaying signal collection until the excitation light had been off for 30 μsec and integrating for a period of 1 msec. The slits of the excitation and emission monochromators were set at 2 nm, and the polymers were mounted on glass plates, cut to fit the cuvette holder of the phosphorimeter. The long luminescence lifetimes exhibited by the lanthanide are compatible with less expensive and relatively slow electronics.

Mixed Ligand Approach

MIPs were prepared for using a mixed ligand approach (**3a** and NTFA) in a matrix of MMA and EGDMA using toluene as a porogen. This approach was required since the vinyl-substituted ligands made weaker complexes than their unsubstituted counterparts. Washing the mixed ligand polymer with methanol and acetone resulted in loss of significant amounts of europium. This is a result of the statistical nature of the ligand substitution such that some of the imprint complex is not bound to the polymer. When the polymers were applied to rebinding studies, the result was a loss in luminescence as the PMP was added

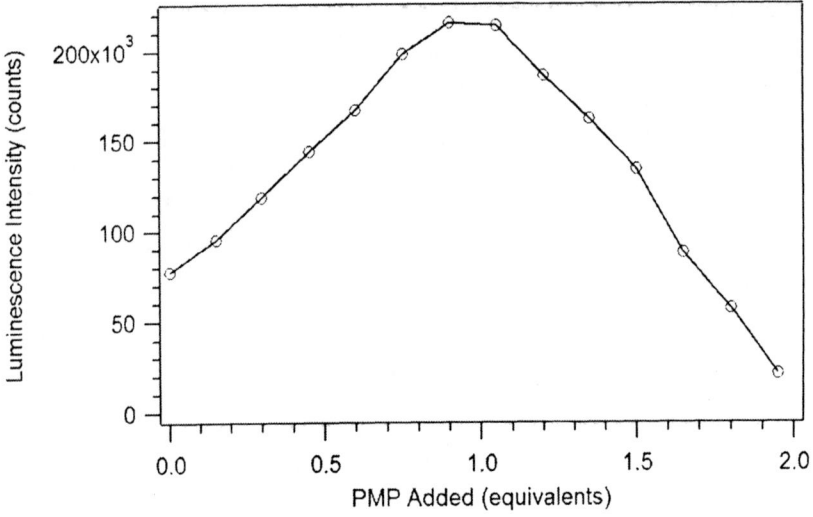

Figure 9. Plot of the intensity of the 611 nm luminescence peak versus equivalents of PMP added.

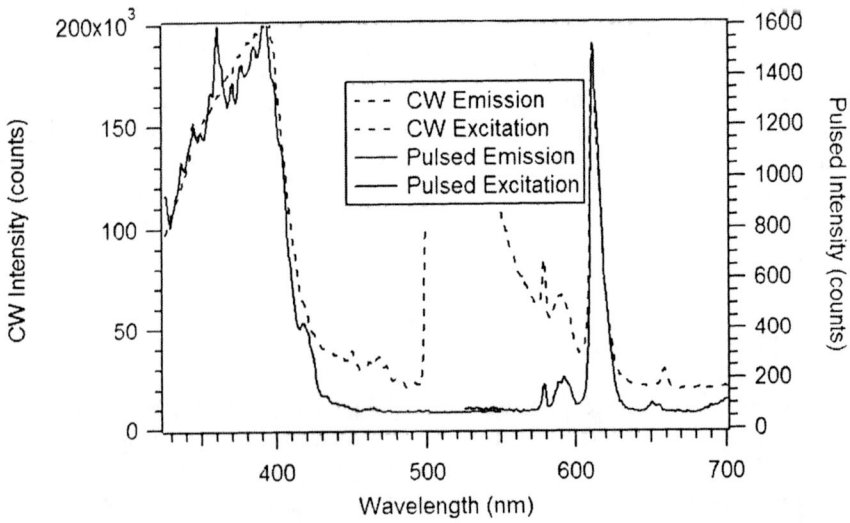

Figure 10. The use of time resolved luminescence to discriminate against background luminescence and scattered light.

This was interpreted as dissociating the complex, due to the weaker interaction with the vinyl-substituted ligand. This approach was viewed as unsatisfactory so a different approach was followed.

RAFT Polymers

Polymers were prepared by the RAFT polymerization method as described above. Initial cleaning was attempted using three refluxing solvents, acetone, methanol and isopropanol. Isopropanol had the highest concentration of PMP after cleaning and was used in all subsequent cases. In a first study, 5 mg of the isopropanol-cleaned polymer were placed in a cuvette in xylene and examined by luminescence spectroscopy. Ten minute intervals were used between additions of PMP. A calibration curve generated by this process is presented as Figure 11. The curvature of the titration plot suggests that either there were insufficient sites in the polymer to cover the dynamic range or that ten minute intervals were not long enough to reach equilibrium.

In order to assure complete removal of PMP, the polymer was next cleaned by Soxhlet extraction with isopropanol. The concentration of PMP in the solvent was measured hourly until a steady state concentration was achieved. Extraction of the RAFT polymers does not liberate any europium as determined by fluorescence of the washings, showing that the complexes are fully incorporated into the polymer. Since the incorporation of the europium complex is better for the RAFT polymer than the mixed ligand polymer, smaller amounts of polymer can be used to get good sensitivity and dynamic range. A second titration curve was generated using the RAFT polymer and longer equilibration times in order to ascertain whether equilibrium is slow (site accessibility is kinetically hindered) or if the cleaning was inadequate and too few sites were available. A control solution of the model compound, $Eu(NTFA)_3$ was used to adjust for instrument changes, resulting in better linearity (Figure 12).

The slow kinetics associated with the first RAFT polymers was suspected to be caused by a lack of porosity or surface roughness. This hypothesis was tested by the use of a scanning electron microscope to investigate the particle's surfaces. As seen in Figure 13, the mixed ligand polymer particles have a rough surface, which is conducive to fairly rapid equilibration. Figure 14 shows that the surface of the RAFT polymer is much smoother. This effect is attributed to the much more even and controlled reaction of RAFT polymerization. In order to improve the surface roughness and porosity, we changed the porogen/solvent to methoxyethanol, a highly polar solvent, to induce phase separation during polymerization, Figure 15. The particles from the new solvent were smaller. The RAFT polymer prepared using methoxyethanol was observed to form an opaque light pink polymer that readily cracked when solvent was removed. The polymer ground to a very fine powder and appeared to completely dissolve in

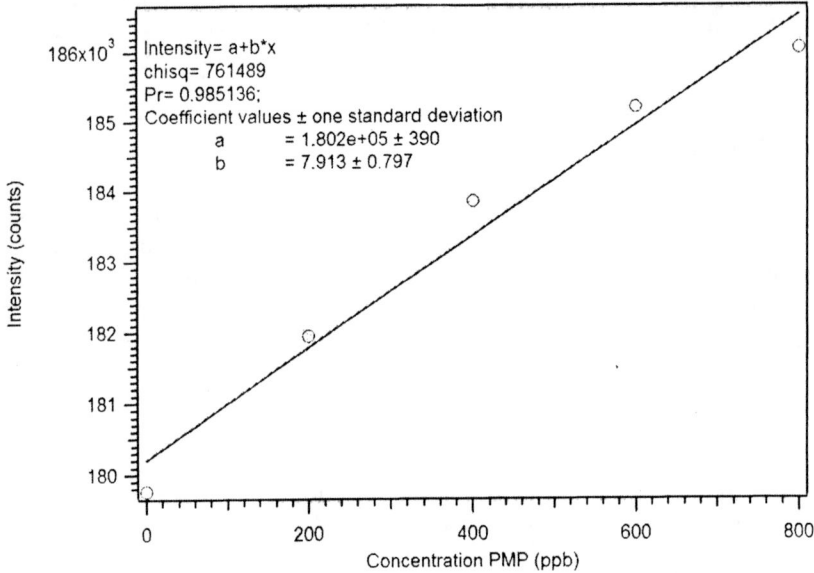

Figure 11. Calibration curve for the RAFT PMP MIP.

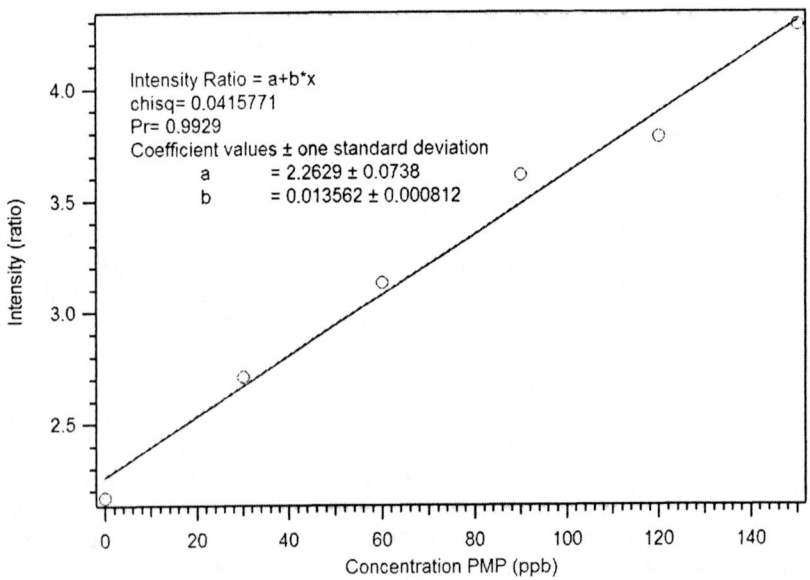

Figure 12. Calibration curve using longer time intervals and an I/Io ratio to correct for instrument changes showing much improved sensitivity.

Figure 13. Scanning electron micrographs of a large particle of the mixed ligand polymer (left 1000 X, right 3000 X). (Reproduced with permission from reference 20. Copyright 2007 Elsevier.)

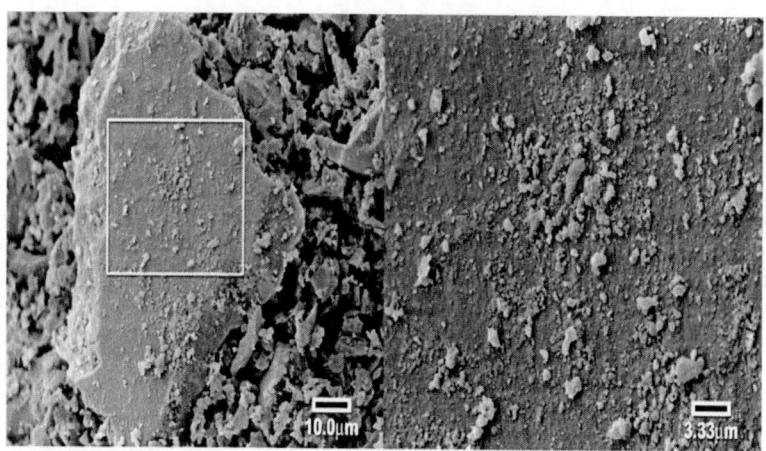

Figure 14. Scanning electron micrograph of a large particle of the RAFT polymer using toluene as the solvent (left 1000 X, right 3000 X). (Reproduced with permission from reference 20. Copyright 2007 Elsevier.)

Figure 15. Scanning electron micrograph of a large particle of the RAFT polymer using methoxyethanol as the solvent (left 1000 X, right 3000 X). (Reproduced with permission from reference 20. Copyright 2007 Elsevier.)

xylene, showing that the refractive indices were matched, which mitigated scattering losses during the fluorescence titrations.

Conclusion

Molecularly imprinted polymers with good sensitivity (limits of detection in the low parts-per-billion range) and very high selectivity (no interference from near identical interferents) have been prepared using RAFT polymerization. The polymers were designed to be used in a liquid light guide based sensor system. The polymers have the same refractive index as xylene and both the polymer and xylene should work well in a Teflon AF light guide. Polymers prepared using methoxyethanol have better porosity and exchange kinetics than do polymers prepared using toluene as solvent and porogen. New, soluble, star polymers (22) may mitigate any remaining issues present in this approach to sensing organophosphorus compounds.

References

1. Wulff, G.; Sarhan, A. *Angew. Chem., Int. Ed.* **1972**, *11*, 341-346.
2. Kriz, D.; Ramstrom, O.; Mosbach, K. *Anal. Chem.* **1997**, *69*, 345A-348A.
3. Murray, G. M.; Jenkins, A. L.; Bzhelyansky, A.; Uy, O. M. *JHUAPL Tech. Digest* **1997**, *18*, 432-441.

4. Arnold, B. R.; Jenkins, A. L.; Uy, O. M.; Murray, G. M. *JHUAPL Tech. Digest* **1999**, *20* 190-198.
5. Jenkins, A. L.; Murray, G. M. *Anal. Chem.* **1996**, *68*, 2974-2980.
6. Jenkins, A. L.; Uy, O. M.; Murray, G. M. *Anal. Chem.* **1999**, *71*, 373-378.
7. Murray, G. M.; Uy, O. M. In *Molecularly Imprinted Polymers*; Sellergren, B., Ed.; Elsevier: Amsterdam, 2000, pp 441-465.
8. Shea, K. J.; Stoddard, G. J. *Macromolecules* **1991**, *24*, 1207-1209.
9. Dhar, D. N.; Lal, J. B. *J. Org. Chem.* **1958**, *23*, 1159-1161.
10. Bradsher, C.; Brown, F.; Blue, W. *J. Am. Chem. Soc.* **1949**, *71*, S3570-3570.
11. Hammond, G. S.; Borduin, W. G.; Guter, G. A. *J. Am. Chem. Soc.* **1959**, *81*, 4682-4686.
12. Morin, M.; Bador, R.; Gechaud, H. *Anal. Chem.* **1989**, *219*, 66-77.
13. Boyd, J. W.; Cobb, G. P.; Southard, G. E.; Murray, G. M. *JHUAPL Technical Digest* **2004**, *25*, 44-49.
14. Chiefari, J.; Chong, Y. K.; Ercole, F.; Krstina, J.; Jeffery, J; Le, T. P. T.; Mayadunne, R. T. A; Meijs, G. F.; Moad, C. L.; Moad, G.; Rizzardo, E.; Thang S. H. *Macromolecules* **1998**, *31*, 5559-5562.
15. Chen, M.; Ghiggino, K. P.; Mau, A. W. H; Rizzardo, E.; Sasse, W. H. F.; Thang, S. H.; Wilson, G. J. *Macromolecules* **2004**, *37*, 5479-5481.
16. Zhou, G.; Harruna I. I. *Macromolecules* **2004** *37*, 7132-7139.
17. Barner-Kowollik, C.; Davis, T. P.; Heuts, J. P. A.; Stenzel, M. H.; Vana, P.; Whittaker, M. *J. Polymer Science Part A: Polymer Chem.* **2003**, *41*, 365-375.
18. Mayadunne, R. T. A.; Jeffery, J.; Moad, G.; Rizzardo, E. *Macromolecules* **2003**, *36*, 1505-1513.
19. Southard, G. E.; Murray, G. M. *J. Org. Chem.* **2005**, *70(22)*, 9036-9039.
20. Southard, G. E.; Van Houten, K. A; Murray, G. M. *Anal. Chimica Acta*, **2007**, *581*, 202-207.
21. Brittain, H. G. *Inorg. Chem.* **1980** 19, 640-643.
22. Southard, G. E.; Van Houten, K. A.; Murray, G. M. *Macromolecules*, **2007**, in press.

Chapter 3

Development of an Enzyme-Based Photoluminescent Porous Silicon Detector for Chemical Warfare Agents

Bradley R. Hart[1], Sonia E. Létant[1], Staci R. Kane[1], Masood Z. Hadi[2], Sharon J. Shields[1], Tu-Chen Cheng[3], Vipin K. Rastogi[3], J. Del Eckels[1], and John G. Reynolds[1,*]

[1]Forensic Science Center, University of California, Lawrence Livermore National Laboratory, Livermore, CA 94551
[2]Lockheed Martin Corporation, Sandia National Laboratory, Livermore, CA 94551
[3]Edgewood Chemical and Biological Center, U.S. Army, Aberdeen Proving Ground, Edgewood, MD 21010

Photoluminescent (PL) porous silicon (PSi) has been modified for the attachment of biomolecules. Silicon wafers were etched by HF in CH_3CH_2OH and current to yield N-type PSi surfaces. A linker was constructed using hydrosilation reactions to give a direct Si-C bond on the surface. The other end of linker was designed so traditional protein cross-linking chemistry could be used to attach biomolecules. Dansyl cadaverine and Biotin-Streptavidin were attached to verify the utility of, and substantiate spectroscopically, the linking system. Glucuronidase was immobilized on the surface utilizing the linking system. The enzyme retained activity monitored through the conversion of p-nitro-phenyl-β-D-glucuronide to p-nitro-phenol.

The photoluminescence of the surface was retained, but varied upon enzymatic production of the p-nitro-phenol. *Alteromonas* sp. JD6.5 (OPAA-2) was also immobilized on the surface and exhibited sufficient enzymatic activity to transform *p*-nitro-phenyl-soman to *p*-nitro-phenol. The change in the photoluminescence of the surface was correlated to enzymatic activity producing the *p*-nitro-phenol hydrolysis product.

Introduction

The use of chemical threat detector systems is critical in the fight against terrorism and in establishing homeland defense (*1*). Although commercial systems are available, development of new detection systems is still required because of the need for broad-based monitoring of large varieties of chemical species as well as the need for better specificity and selectivity towards individual very toxic species (*2*). Applications range from personnel protection, such as in first responders, to facility protection, such as airports, subways and office buildings. The chemical threats range from simple industrial chemicals, such as HCl, with moderate toxicity, to complex organophosphorus nerve agents, such as sarin, with extreme toxicity.

Of the organophosphorus nerve argents, the G agents are particularly important because of extreme toxicities that comes from their cholinesterase activity, coupled with their high volatility, in some cases, making them viable inhalation threats (*3*). Detection systems for these agents must be highly sensitive and quick responding because of their acute and fast toxicity.

Recently, we have been attempting to develop chemical weapons detection systems using immobilized enzymes on the surface of PL PSi (*4–6*). The approach is to chemically treat silicon to be porous and PL, construct a linker system that attaches to the PSi surface through a direct Si-C bond, covalently attach an enzyme active towards the target chemical compound, and monitor the enzymatic activity through the PSi photoluminescence. The ultimate aim is to develop a generalized methodology that can be used to detect any target compound, if the appropriate enzyme can be attached. Various mechanisms could change the PL, such as quenching due to product formation and enzyme conformational changes. The preferred mechanism would relate to changes occurring in the enzyme during the action of enzymatic chemistry. Such has been noted in OP enzymes encapsulated into quantum dots (*7*).

Here we report the development of the methodology employed in attaching enzymes and measuring chemical changes occurring when selected attached enzymes are active towards example substrates. PL changes have been tentatively identified occurring due to reversible quenching due to charge transfer interactions with one of the substrate breakdown products.

Experimental

The techniques used here have been published in detail elsewhere (*4–6*). It is important to reiterate that one of the chemical warfare agent surrogates used is *p*-nitro-phenyl-soman. It should be pointed out that, to our knowledge, there is no published data for the human toxicity of the substrate *p*-nitro-phenyl-soman. It, therefore, has to be assumed to be comparable to the available data for soman (the LD_{50} values for a 70 kg person are 350 mg for skin exposure and 30 mg for ingestion) (*8*). Safety measures include preparing all the substrate dilutions in a fume hood while wearing proper personal protective equipment and using a perfectly sealed flow cell.

PNS
p-nitro-phenyl-soman

p-nitro-phenyl soman was selected as the surrogate because of several reasons: it is a well characterized material that has some similar properties to soman; it is less volatile than soman, so it is easier to handle; it has better water stability than soman, so hydrolysis is not as much of an issue; the reaction with OPAA yields *p*-nitro-phenol and has been well characterized for the non-immobilized enzyme, and it is a compound that can be monitored by UV-vis spectroscopy (*9*).

The flow cell was designed in-house. The main body was constructed out of Teflon with quartz windows and Teflon gaskets. The system was tested thoroughly with non-toxic materials to assure leak-free flow through.

Results and Discussion

Photoluminescent Porous Silicon

PL properties of PSi have been known for over a decade (*10*). During that time, many different etching techniques have been developed to give a wide variation in porosity and optical properties. For this application, a material with intense photoluminescence and high density of surface hydrides is desired. In the approach described here, the PSi platform is fabricated from Si wafers etched in HF/C_2H_5OH with current. Both *N* and *P*-type PSi provide reactive H terminated Si surface sites for attaching the linker system. Figure 1 shows the scanning electron micrograph (SEM) and the emission spectra for both *N* and *P* type etched surfaces. Both have similar etching efficiency (consumed hydride). The *N*-type is less transparent in the IR, which makes surface characterization more difficult. However, the majority of experiments were performed using *N*-type because of higher PL intensity.

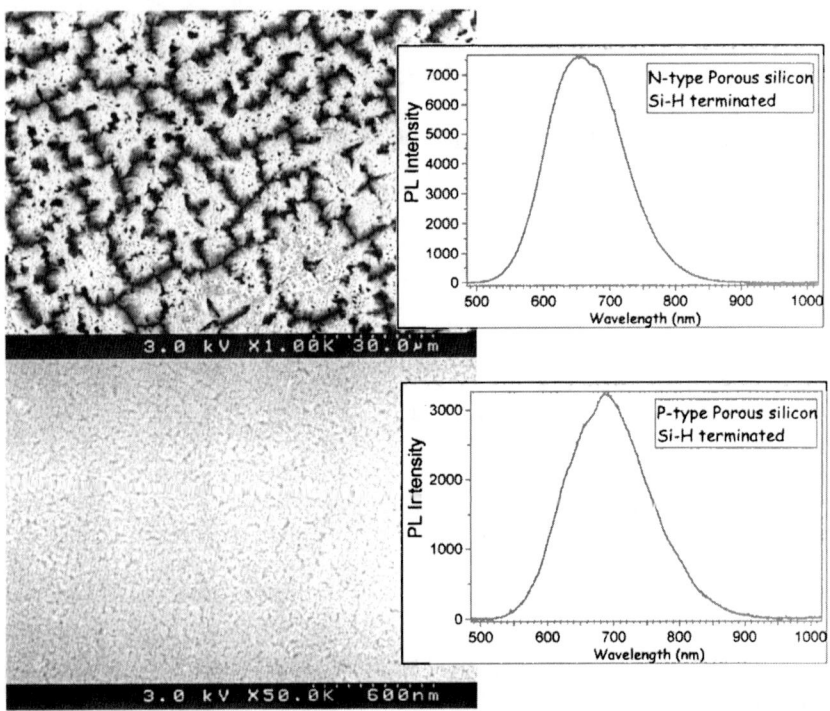

Figure 1. Modified PSi surfaces derived from different etching methods examined for use as the platform for the OPAA enzyme. The left side features are scanning electron micrographs (SEM) with corresponding emission spectra at 290 nm excitation (λ_{ex} = 290 nm) on the right side for N-type PSi and P-type PSi. The N-type PSi has a much larger pores as well as a much high PL than the P-type, and was selected for most of the surface functionalization. (Reproduced with permission from reference 4. Copyright 2003 The Royal Society of Chemistry.)

Linker System

Chemically active linking systems can be connected to the PSi surface by a variety of methods, such as Lewis Acid hydrosilation, reaction with siloxanes, photo-assisted hydrosilation, and others (*11–17*), which attach the linker to the surface through either a Si-O or a Si-C moiety. The oxides introduce surface electronic defects that may limit the potential of using PL for sensing, suggesting the Si-C linking site may have advantages (*12*). In addition, attachments with the Si-C bond appear to form a more resilient and uniform surface attachment than attachments with through the oxide layer (*12*).

Figure 2 shows the synthetic steps to form the linker system by the Lewis Acid hydrosilation route (SPDP = *N*-succinimidyl-3-(2-pyridyldithio) propionate, DTT = dithiothrietol, GMBS = *N*-(γ-maleimidobutyryloxy)succinimide ester), BOC = *t*-butyloxycarbonyl, TFA = trifluoroacetic acid). After the PSi surface is prepared as above, the surface is washed with a dilute solution of HF to remove incidental oxidation that might have occurred during storage. The Si-C bond is formed by the reaction of propeneyl nitrile with the surface catalyzed by $EtAlCl_2$. The nitrile is then reduced to 1° amine using $LiAlH_4$. At this point, construction of the linker is through traditional protein cross-linking chemistry at the amine. SPDP is attached to the 1° amine through the NHS-activated ester providing a 2-mercaptopyridine protected thiol. This thiol is deprotected by reductive cleavage using DTT, and then is reacted selectively with maleimide of GMBS, providing an NHS-activated ester as the site of attachment for the biomolecule.

The first step of the synthesis can be accomplished through the alternate pathway (not shown) of light promoted hydrosilation of a BOC protected butyleneamine (*6*). The protection is then cleaved using TFA. At this point, the balance of the linker formation chemistry is the same as in the first route, yielding the NHS-activated ester as the site for attachment. In this case, instead of a Si bonded to a sp^2 C, as in the case of the Lewis Acid hydrosilation method, the Si is bonded to a sp^3 C.

Both synthesis routes perform well to produce a useable linker. We have used both methods extensively in our studies. However, there are synthetic advantages in using the light assisted method. In particular, the first step is a little easier to execute and removes the need for $EtAlCl_2$.

Characterization of Functionalized PSi Surface

With each reaction building the linker system, the surface was characterized by IR. Figure 3a shows an example of the surface before and after the Lewis Acid hydrosilation step. The surface before hydrosilation shows prominent features at 2100 cm^{-1} and 1100 cm^{-1} assigned to Si-H and Si-O bonds. Upon Lewis acid hydrosilation, additional features at 2900 cm^{-1}, 2200 cm^{-1}, 1500 cm^{-1} and 1300 cm^{-1}, assigned to C-H, C triple bonded to N, C=C, and CH_2, respectively, establishing the presence of the attached alkene to the surface, are present. The sharp feature at 950 cm^{-1} is due probably due to SiH_2 (*18*). Note, however, there still is some intensity due to Si-O, indicating that the surface still has oxidation, as well as there is still significant intensity due to Si-H, indicating

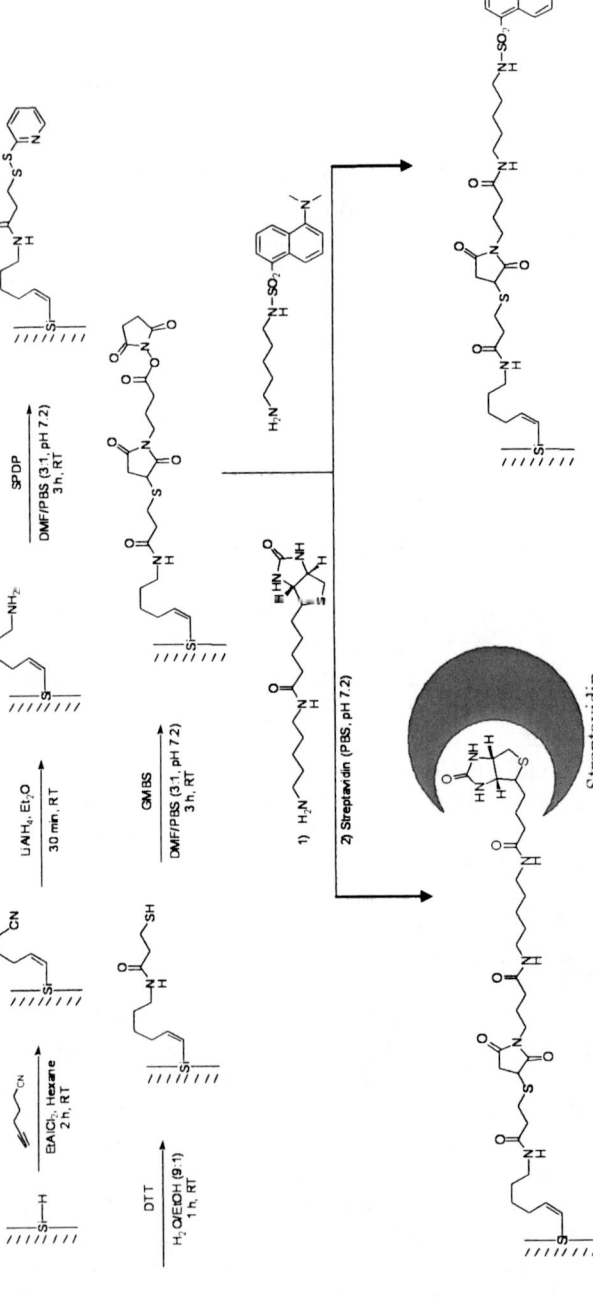

Figure 2. Immobilization linker system constructed using the Lewis Acid hydrosilation route (SPDP = N-succinimidyl-3-(2-pyridyldithio) propionate, DTT = dithiothrietol, GMBS = N-(γ-maleimidobutyryloxy)succinimide ester).

that not all of the surface Si-H sites reacted to form the base of the linker. By comparing IR intensities of the two samples, this corresponds roughly to 25% of the Si-H groups reacted.

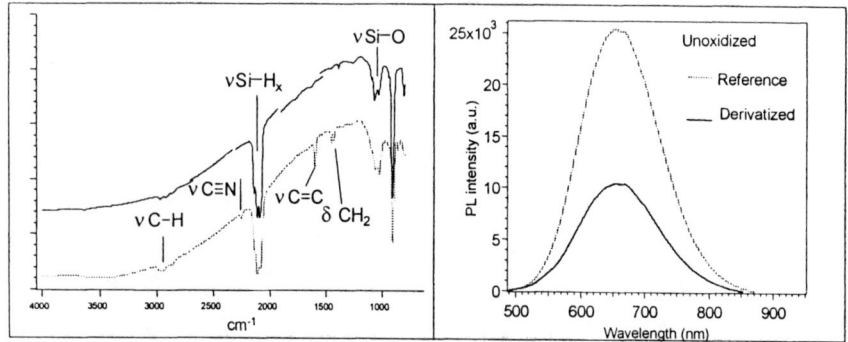

Figure 3. FTIR (a) and fluorescence (b) spectra of PSi surface before and after Lewis Acid hydrosilation reaction (first step) to produce the Si sites for linker construction. The appearance of the C-H (2900 cm^{-1}), nitrile (2200 cm^{-1}), C=C (1500 cm^{-1}) and CH_2 (1300 cm^{-1}) indicate the presence of the success of the hydrosilation reaction.

Figure 3b shows fluorescence spectra of the same two samples at λ_{ex} = 290 nm. The surface before Lewis Acid hydrosilation shows an emission intensity of about 25 x 10^3 a.u., while after the hydrosilation, the maximum has decreased to about 10 x 10^3 a.u. In all cases measured, this decrease was typical after the first synthesis step. Subsequent synthesis steps did not reduce the emission intensity further. This may be due to oxidation that occurs in the first derivatization step that does not increase upon further reaction steps.

To estimate surface coverage of the linker, the protected thiol formed in the SPDP reaction was cleaved using DTT. This produces 2-mercaptopyridine in stoichiometric amounts related to the bound linker. The quantity of 2-mercaptopyridine was measured by UV-vis spectroscopy resulting in roughly 2.5 x 10^{-9} molar linker concentration for the Lewis Acid catalyzed hydrosilation method and 3 times that value for the light assisted hydrosilation method.

To test the ability of the linker system to attach molecules to the PSi surface, dansyl cadaverine was reacted with the NHS activated ester group on the surface as shown in the last synthesis step in Figure 2. The fluorescence of the surface was then examined at λ_{ex} = 325 nm. The emission spectrum, Figure 4, exhibits a dominant emission around 680 nm, due to the PSi surface, as seen in Figure 3b, and the presence of emission at 518 nm due to the presence of dansyl group. Note the luminescence remains high.

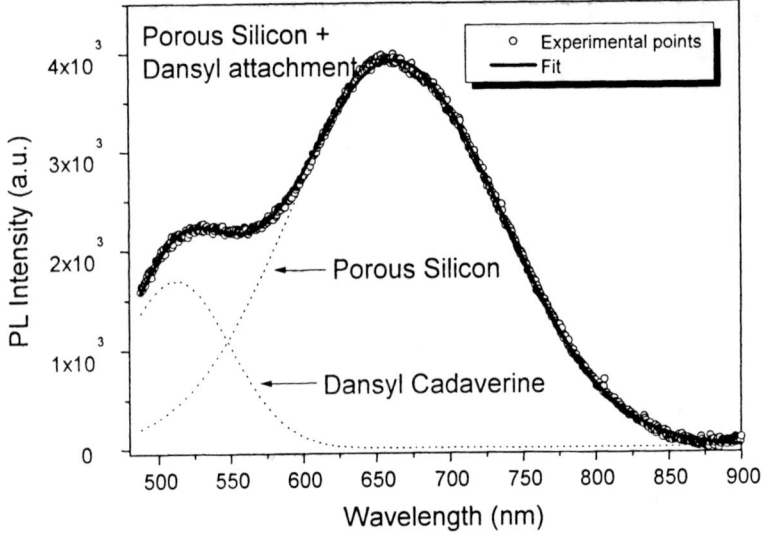

Figure 4. Emission spectra of PSi surface that has been functionalized with dansyl dye attached. The dansyl group is evident by the formation of the shoulder at about 518 nm in the emission spectrum. (Reproduced with permission from reference 4. Copyright 2003 The Royal Society of Chemistry.)

The ability of the linker system to attach large biomolecules to the surface was also explored using Biotin-Streptavidin. In this case, the Biotin was covalently attached through the NHS activated ester, as seen in last synthesis step of Figure 2. Streptavidin, which has a very high affinity for Biotin, was then reacted with the surface. After thorough washing to eliminate non-specifically absorbed biomolecules, the surface was exposed to Trypsin digest and the protein fragments were then examined by mass spectrometry. Figure 5 shows the MALDI-TOF-MS of the digest. Using the protein digest database, the presence of Streptavidin is confirmed by identifying key peptide fragments after digestion.

Development of a Flow Cell

In the design of the detector system, the photoluminescence should be measured during enzymatic activity. This required the construction of a flow cell system that could be interfaced with both a UV-vis and fluorescence spectrometer. In addition, because ultimately the system will be handling very toxic materials, the complete system must be sealed to protect the operators.

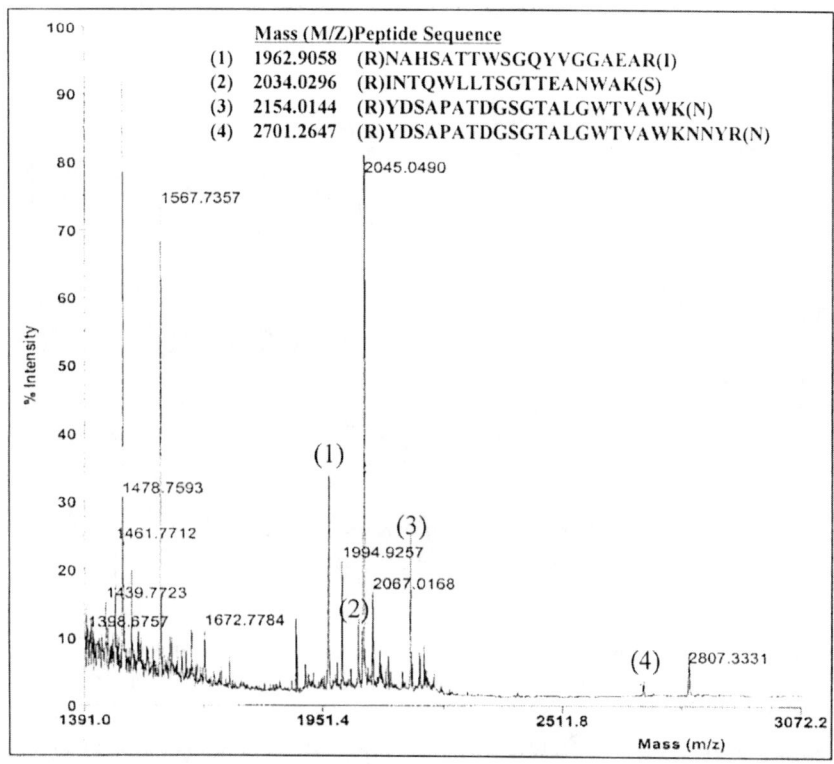

Figure 5. MALDI-TOF-MS of Trypsin digest of PSi surface functionalized with Biotin-Streptavidin attached. The proteins shown correspond to the Trypsin digest of Streptavidin indicating that Streptavidin was attached to the Biotin.

Such a system, constructed with corrosion resistant materials, is shown in Figure 6. The functionalized PSi surface with the enzyme attached is placed into the Teflon cell, shown in the right hand side of Figure 6. The cell has a window that allows for the fluorescence interrogation and detection through optical fibers of the PSi surface. The flow of the system is controlled by a closed loop of tubing from a peristaltic pump to the flow cell, to a closed cuvette in UV-vis spectrophotometer. Addition to the closed system is through the injection port on the flow cell.

Attachment of Glucuronidase

To test the concept of attaching an enzyme to the surface, Glucuronidase (GUC) was attached to the functionalized PSi surface. Figure 7 shows an SEM

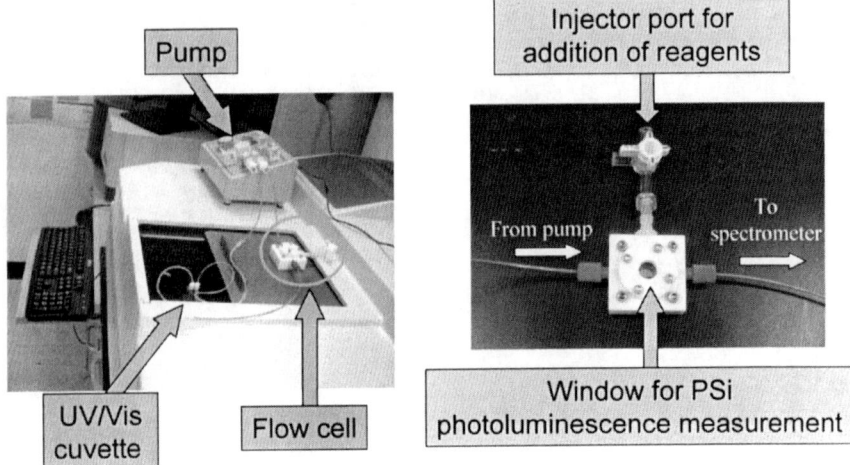

Figure 6. Sealed flow cell system for real-time fluorescence detection of PSi platform with enzymes attached.

of the PSi surface that has gone through enzyme immobilization. The linker system (a) is described from the light promoted hydrosilation method and the site for binding is not implied. The top view (b) shows the etching produces features that are around 750 nm in diameter. The side view (c) shows the porous etching extends about 25 to 30 μm into the surface.

The enzymatic activity of the GUC immobilized on the PSi surface was tested in the flow system described above. The activity was measured by monitoring, by UV-vis spectroscopy, the formation of products of the enzymatic action. GUC is an enzyme that converts p-nitro-phenyl-β-D-glucuronide to p-nitro-phenol. The enzymatic chemistry has been well characterized for the solution phase enzyme (19).

Figure 8 shows the activity of the immobilized GUC towards various concentrations of p-nitro-phenyl-β-D-glucuronide in the flow cell system described in Figure 6. The inset of the figure shows the behavior of the formation of the enzymatic degradation product, p-nitro-phenol with time, monitored by UV-vis spectroscopy at 405 nm, upon exposure of the immobilized enzyme to different concentrations of the substrate, 25, 50, 75, 125, and 250 μM. The build-up of product increases linearly with time, with the slope of this linear relationship increasing with increasing initial concentration of the substrate. The slopes of these lines are then used to form the rate (V) vs. substrate concentration graph shown in Figure 8.

The data were then analyzed using the Michaelis-Menten equation (19). The Michaelis-Menten equation is derived from a simple kinetic model for a

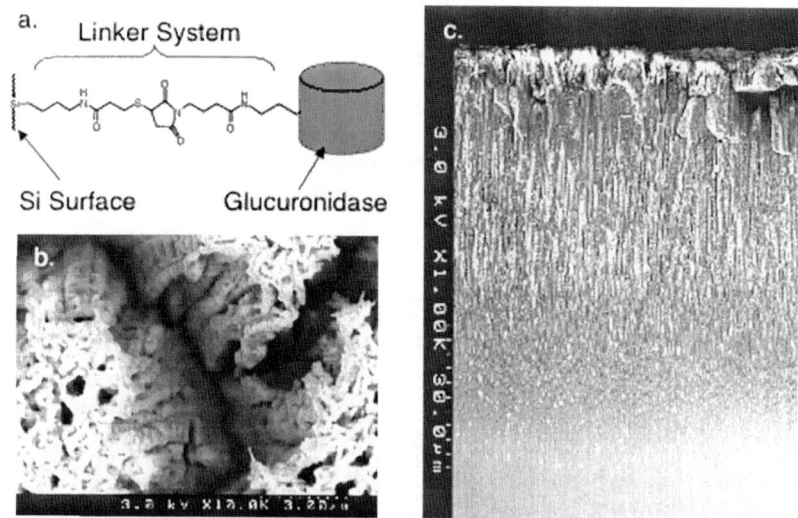

Figure 7. GUC fastened to functionalized PSi surface, artist conception (a) functionalized surface with enzyme attached where the binding site on the GUC is not specified, scanning electron micrographs of the surface from a top view (b) and side view (c). (Reproduced with permission from reference 5. Copyright 2004 Wiley-VCH.)

single-substrate, non-cooperative, enzyme-catalyzed reaction that accounts for the isotherm relationship between substrate concentration and reaction rate. V is the initial velocity, or rate, of the reaction, K_m is the Michaelis constant and V_{max} is the maximum rate approached by very high concentrations of substrate.

The curve in Figure 8 is the equation and the points are the data derived from the reduction of the data in the inset of the figure. The activity constants derived from the Michaelis-Menten equation are V_{max} = 0.11 nmol·min^{-1} and K_m = 0.1 mM. The spectrophotometric assay values of GUS in solution are V_{max} = 1.55 nmol·min^{-1}, and K_m = 0.078 mM. The velocity maximum value for the immobilized enzyme is about 1/10th the value of the solution assay value (*19*), probably due to the immobilization in the PSi surface framework restricting access to the enzymatic sites, effectively reducing the concentration of the available enzyme. The K_m values, however, are similar, indicating the activity of the enzyme is not affected by the immobilization.

Figure 9 shows the fluorescence behavior of the immobilized enzyme in the above experiment. As shown in the flow cell system in Figure 6, the PSi surface is excited at 290 nm and emission monitored at 560 nm. The product build-up was also measured by UV-vis at 405 nm. Examining the response of the PL

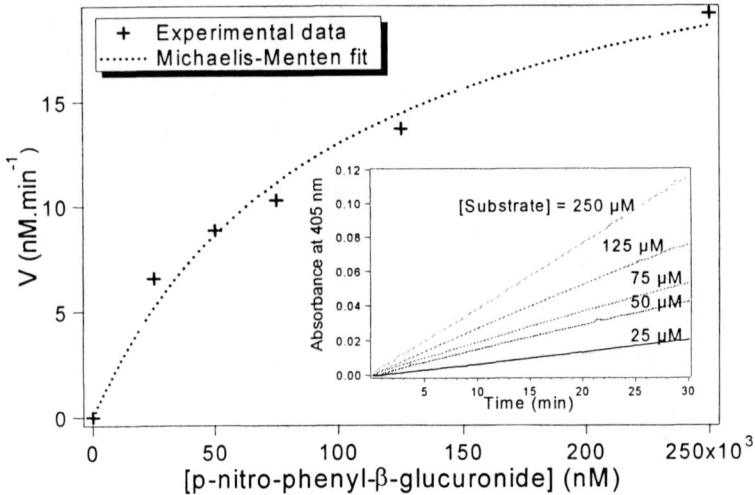

Figure 8. Activity of GUC immobilized on PSi towards p-nitro-phenyl-β-D-glucuronide. Crosses are data, curve is the Michaelis-Menten equation. (Reproduced with permission from reference 5. Copyright 2004 Wiley-VCH.)

with time for the 25 μM substrate concentration shows that emission intensity decreases linearly with time. This is true for all the substrate concentrations shown, as well as the effect increases in magnitude with increase substrate concentration. Concurrent with this, the product build-up increased linearly.

PL lifetimes were also examined at product concentrations and are shown in Figure 10. In pure buffer, the lifetime was measured to be 10 μsec and, in the presence of *p*-nitro-phenol, the lifetime was measured to be 6 μsec. This decrease indicates the presence of *p*-nitro-phenol partially quenching the fluorescence. Combining this observation with the PL reduction correlated with *p*-nitro-phenol build-up, as shown in Figures 9 and 10, indicates the PL is partially quenched by a reversible mechanism, most likely charge transfer between the *p*-nitro-phenol and the PSi surface.

Attachment of Alteromonas sp. JD6.5 OPAA-2

The ultimate goal of this project is to design enzyme-based PL PSi detectors for toxic chemical detection; therefore, the selection of enzymes active towards these types of target molecules is critical. Table I shows activity parameters of some enzymes in solution that have hydrolysis activity toward G-type nerve agents. OPAA-2 was selected for immobilization because it has relatively high activity towards GB and GD, as well as towards surrogates such as DFP, and *p*-nitro-phenyl-soman *(20–23)*. The products from the enzymatic action would be a phosphonic acid and HF.

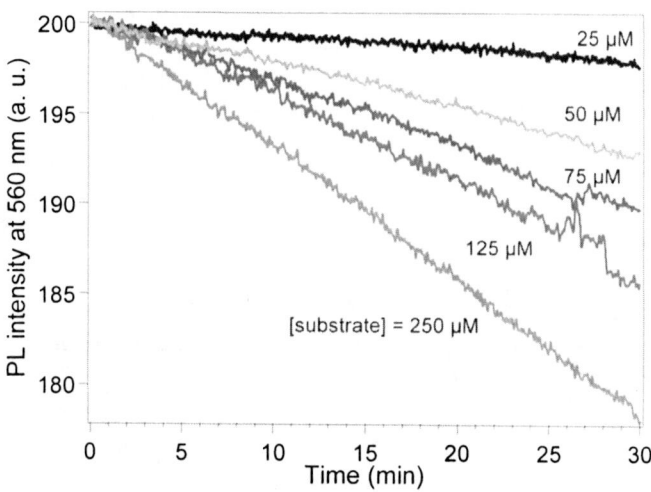

Figure 9. PL monitored at 560 nm of PSi with GUS immobilized on the surface. Formation of p-nitro-phenol was monitored at 405 nm. (Reproduced with permission from reference 5. Copyright 2004 Wiley-VCH.)

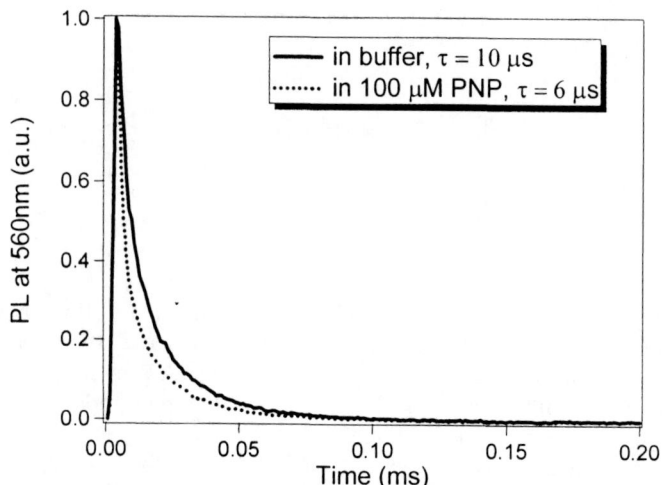

Figure 10. Fluorescence lifetimes for the PL with and without the formation of PNP. (Reproduced with permission from reference 5. Copyright 2004 Wiley-VCH.)

Table I. Hydrolytic catalytic activity, K_{cat} (sec^{-1}), of selected enzymes toward some nerve agents

Organism	Enzyme	GA	GB	GD	DFP	VX
Pseudomonas diminuta	OPH	ND	56	5	465	0.3
Alteromonas sp. JD6.5	OPAA-2	94	611	3145	650	0
Alteromonas haloplanktis	ND	111	257	1389	575	0
Alteromonas undina	ND	300	376	2496	239	0

NOTE: ND = not determined, data from Reference 24.

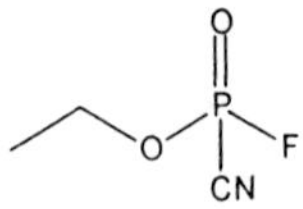

GA (Tabun)

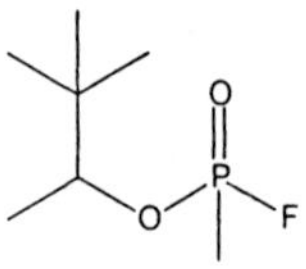

GD (Soman)

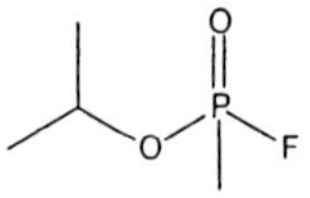

GB (Sarin)

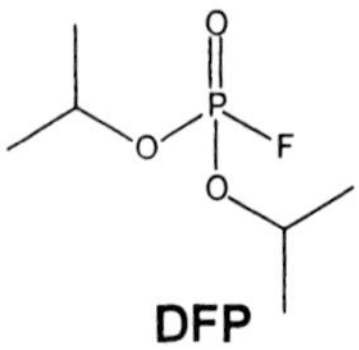

DFP

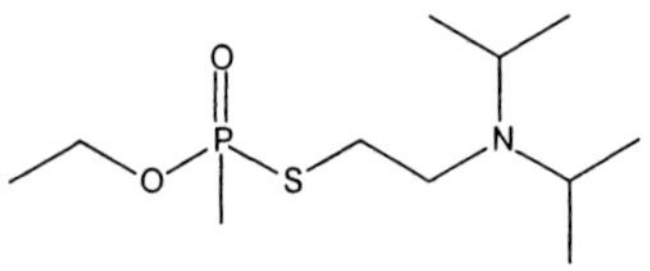

VX

Figure 11 shows p-nitro-phenol production as a function of time by OPAA-functionalized PSi device recorded in the flow cell in real time for various concentrations of substrate p-nitro-phenyl-soman. OPAA-2 hydrolyzes p-nitro-phenyl-soman to p-nitro-phenol and pinacoyl methyl phosphonic acid. The production of p-nitro-phenol is monitored at 405 nm. The p-nitro-phenol concentration increases with time. In addition, as the starting substrate concentration increases, at a specific time, the more p-nitro-phenol is produced. This behavior is very similar to the enzyme in solution (not shown), except the amount of p-nitro-phenol product is much less due to probable enzyme access issues as discussed for GUS immobilization. These results indicate that OPAA-2 and potentially other OPH-type enzymes can be successfully immobilized with the linking systems described here while retaining enzymatic activity.

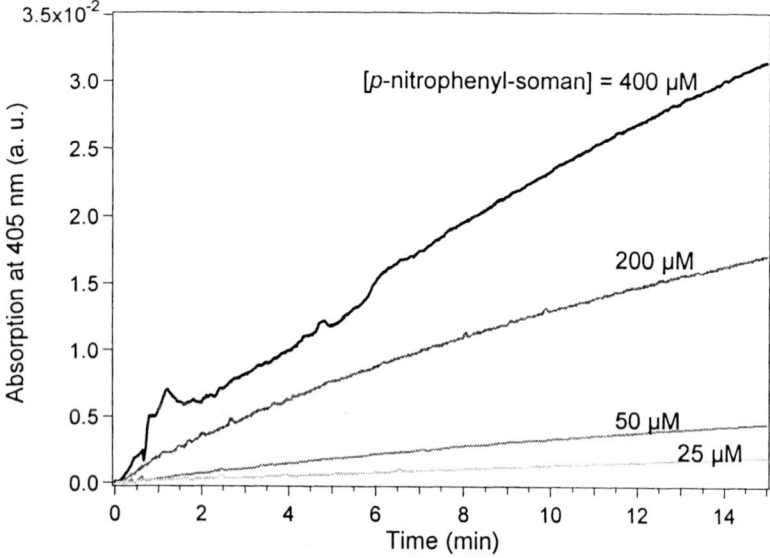

Figure 11. Detection of p-nitro-phenol production as a function of time by OPAA-functionalized PSi device recorded in the flow cell in real time for various concentrations of p-nitro-phenyl-soman. p-nitro-phenol production is measured by the appearance of the 405 nm UV-vis absorption band. (Reproduced with permission from reference 6. Copyright 2005 The Royal Society of Chemistry.)

Figure 12 shows the emission spectra (λ_{ex} = 290 nm) of OPAA-functionalized PSi prior to exposure to *p*-nitro-phenyl-soman, during exposure to 25 µM and 50 µM substrate concentrations, and following PBS buffer washes. The black lines depict the photoluminescence signal of a functionalized sample in bis-trispropane buffer solution before and after the injection of substrate and the gray and light gray lines show the photoluminescence signal recorded 5 min after the injection of the substrate. A significant decrease in the photoluminescence associated with the injection of substrate was observed, as seen in the previous work with β-Glucuronidase enzyme (5). There is an inverse linear correlation between the amount of decrease in photoluminescence and the initial substrate concentration. The transduction is fast (less than 2 min), sensitive (10% decrease for 25 µM), and reversible. Moreover, the samples are stable and can therefore be reused.

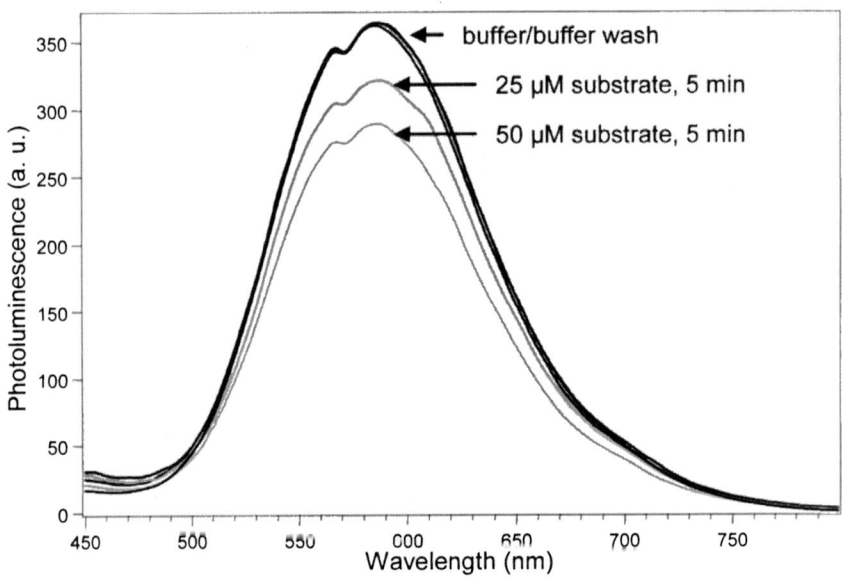

Figure 12. Emission spectra (λ_{ex} = 290 nm) of OPAA-functionalized PSi before exposure to p-nitro-phenyl-soman, during exposure to 25 µM and 50 µM substrate concentrations, and following PBS (phosphate buffered saline) buffer washes. (Reproduced with permission from reference 6. Copyright 2005 The Royal Society of Chemistry.)

Conclusions

PL PSi can be produced that is a suitable platform for enzyme immobilization by variations in normal etching techniques with HF. N-type is preferred due to the higher PL signal. Hydrosilation is a versatile method for surface modification, producing a very stable Si-C bond. The most direct and facile route is photo-initiated. Traditional protein cross-linking chemistry can provide useful functionality for attaching biomolecules to one end of the linker. Significant levels of photoluminescence remain after functionalization. In addition, upon functionalization, the surfaces become resistant to further oxidation. Enzymes, such as GUS and OPAA-2, can be attached and still retain enzymatic activity after attachment. The PSi surface retains PL. Upon product formation from the enzyme action, the PL decreases with increasing product build-up, consistent with a reversible charge transfer quenching mechanism. Glucuronidase has been attached and used to demonstrate proof of principle.

Acknowledgment

This work was performed under the auspices of the U.S. Department of Energy by University of California Lawrence Livermore National Laboratory under contract No. W-7405-Eng-48.

References

1. Reynolds, J. G.; Hart, B. R. *JOM* **2004**, *56*, 36-39.
2. Ellison, D. H. *Handbook of Chemical and Biological Warfare Agents*, CRC Press Boca Raton, **2000**, ISBN 0-8493-2803-9.
3. Institute of Medicine and National Research Council, Chemical and Biological Terrorism, National Academy Press, Washington D.C. **1999**, ISBN 0-309-06195-4.
4. Hart, B. R.; Létant, S. E.; Kane, S. R.; Hadi, M. Z.; Shields, S. J.; Reynolds, J. G. *Chem. Commun.* **2003**, 322.
5. Létant, S. E.; Hart, B. R.; Kane, S. R.; Hadi, M. Z.; Shields, S. J.; Reynolds, J. G. *Adv. Mater.* **2004**, *16*, 689.
6. Létant, S. E.; Hart, B. R.; Kane, S. R.; Hadi, M. Z.; Cheng, T.-C.; Rastogi, V. K.; Reynolds, J. G. *Chem. Commun.* **2005**, 851-853.
7. Ji, X.; Zheng, J.; Xu, J.; Rastogi, V. K.; Cheng, T.-C.; DeFrank. J. J.; Leblanc, R. M. *J. Phys. Chem. B* **2005**, *109*, 3793-3799.
8. *Possible long-term health effects of short-term exposure to chemical agents;* National Academy of Sciences, N. R. C., Committee on Toxicology. National Academy Press: Washington, DC 1982, Vol. 1: anticholinesterases and anticholinergics.
9. Hill, C. M.; Li, W.-S.; Cheng, T.-C.; DeFrank, J. J.; Raushel, R. M. *Bioorganic Chemistry* **2001**, *29*, 27-35.
10. Cullis, A. G.; Canham, L. T.; Calcott, P. D. J. *J. Appl. Phys.* **1997**, *82*(3), 909-965.
11. Buriak, J. M. *Chem. Rev.* **2002**, *102*, 1271-1308.
12. Strother, T.; Cai, W.; Zhao, X.; Hamers, R. J.; Smith, L. M. *J. Amer. Chem. Soc.* **2000**, *122*, 1205-1209.
13. Dancil, K.-P. S.; Greiner, D. P.; Sailor, M. J. *J. Amer. Chem. Soc.* **1999**, *121*, 7925-7930.
14. Wojtyk, J. T. C.; Moran, K. A.; Boukherroub, R.; Wayner, D. D. M. *Langmuir* **2002**, *18*, 6081-6087.

15. Pijanowska, D. G.; Remiszewska, E.; Lysko, J. M.; Jazwinski, J.; Torbicz, W. *Sensors and Actuators B* **2003**, *91*, 152-157.
16. Laurell, T.; Drott, J.; Rosengren, L. *Biosensors and Bioelectronics* **1995**, *10*, 289-299.
17. Subramanian, A.; Kennel, S. J.; Oden, P. I.; Jacobson, K. B.; Woodward, J.; Doktycz, M. J. *Enzyme Microb. Technol.* **1999**, *24*, 26-34.
18. Boukherroub, R.; Morin, S.; Wayner, D. D. M.; Bensebaa, F.; Sproule, G. I.; Barbeau, J.-M.; Lockwood, D. J. *Chem. Mater.* **2001**, *13*, 2002-2011.
19. Aich, S.; Delbacre, L. T. J.; Chen, R. *BioTechniques* **2001**, *30*(4), 846-850.
20. Landis, W. G.; DeFrank, J. J. *Adv. Appl. Biotechnol. Ser.* **1990**, *4*, 183.
21. DeFrank, J. J.; Chen, T.-C. *J. Bacterial.* **1991**, *173*, 1938.
22. Cheng, T.-C.; DeFrank, J. J.; Rastogi, V. K. *Chem. Biol. Interact.* **1999**, *119*, 455-462.
23. Cheng, T.-C.; Rastogi, V. K.; DeFrank, J. J.; Sawiris, G. P. *Ann. N. Y. Acad. Sci.* **1998**, *864*, 253-258.
24. Mulbry, W.; Rainina, E. *Chemistry International* **1999**, *21*(6), 173-178.

Chapter 4

Optical Enzyme-Based Sensors for Reagentless Detection of Chemical Analytes

Brandy Johnson-White[1] and H. James Harmon[2,*]

[1]Center for Bio/Molecular Science and Engineering, Naval Research Laboratory, Washington, DC 20375
[2]Physics Department, Oklahoma State University, Stillwater, OK 74078

As reversible, competitive inhibitors of enzymes, porphyrins can be used for identification and quantification of a substrate or other competitive inhibitors of enzymes. This approach has been successful in the development of cholinesterase and organophosphorus hydrolase bearing glass surfaces for the detection of substrates and inhibitors, including organophosphate compounds. The technique has demonstrated detection limits for sarin (GB) below the Immediately Dangerous to Life or Health levels indicated by CDC/NIOSH.

© 2008 American Chemical Society

Introduction

Organophosphonates (OPs) are a class of compounds derived from phosphoric, or similar, acids, many of which possess anti-cholinesterase activity. Several are used for agricultural applications, including pesticides such as malathion and fenthion, herbicides such as bensulide, and fungicides such as pirimiphos-methyl. The organophosphate sarin (GB) was originally developed in Germany as a pesticide and later used as a nerve agent due to its extreme toxicity to humans (GB is ~1000-fold more toxic than parathion). GB has been implicated in both terrorist attacks and warfare situations (*1*).

Inhibition of cholinesterase activity by OPs results in a build up of neurotransmitters, such as acetylcholine. The constant nerve impulses can cause dizziness, headache, vomiting, suppression of breathing, and can lead to death (*2*). OPs encountered as occupational hazards or in food, water, or air contaminated during legitimate use make detection important for agricultural safety. In addition, the threat of chemical attack by foreign adversaries or terrorist action presents a pressing need for a reliable, real-time sensor for organophosphates.

A key aspect of sensor design and use is specificity. Antibodies (Ab) exhibit selective binding of analytes (antigens, Ag) but may also have cross-reactivity to other antigens, especially small molecular targets. Nonetheless, in competitive binding assays where antibodies bind native antigen or antigen derivatized with a colorimetric/fluorescent indicator or an enzyme that produces a secondary product, small molecules can be readily detected. (Using a secondary enzyme system can "amplify" the response and increase sensitivity by successive enzyme turnovers but at the expense of time.)

Enzymes also possess great binding specificity and, in many cases, the binding affinity of enzyme for its substrate or inhibitor can exceed that of an antibody for an antigen. The commercial availability of a wide variety of enzymes and recent advances in molecular biology allowing for the isolation of the responsible gene(s) and expression of the product in large quantities make enzyme-based technologies attractive. Enzymes can also be modified by genetic manipulation to yield products with potentially more desirable characteristics (*3–6*).

Enzyme-based detection often relies on a change in the rate of production of a measurable product of enzymatic catalysis. Detection of an enzyme substrate is based on comparing the rate of change in product concentration in an unknown concentration of analyte to that in a known concentration of analyte. Detection of an enzyme inhibitor is accomplished using the difference in the rate of change in product concentration in the absence and presence of the inhibitor for a known substrate concentration. The rate of catalysis in enzyme based sensors can be affected by factors such as pH, temperature, or ionic concentration and by loss in enzyme activity and time requirements are imposed by enzyme turnovers required to produce a measurable product concentration.

Porphyrins are a class of aromatic molecules whose intense absorbance and fluorescence characteristics are influenced by their immediate surroundings (*7–9*) and, thus, readily reflect their interaction with other molecules. The large extinction coefficient (500,000 A/cm pathlength/M) of porphyrins allows for measuring small absorbance changes ideal for optical detection applications.

The research presented here combines the sensitivity of the porphyrin macrocycle with the specificity of an enzyme-active site to achieve a selective response to the presence of chemical analytes of interest. The porphyrin is chosen to obtain a particular interaction with the enzyme so that interaction of the porphyrin-enzyme complex with compounds binding at the active site (substrate or competitive inhibitor) results in a change in the porphyrin absorbance (Figure 1). Neither the interaction of the complex with other analytes, including noncompetitive and uncompetitive inhibitors, nor the rate of the enzyme-catalyzed reaction affects the measurement as the analyte binding is directly detected based on the change in the porphyrin absorbance spectrum. Small fluctuations in the enzyme population do not alter the sensor response as detection is based on a change in the absorbance spectrum as compared to the pre-exposure spectrum. This method has been applied for the detection of a variety of analytes using several different enzymes with various porphyrins (*10–18*).

Methods

The measurement technique used for these experiments is evanescent wave absorbance spectroscopy (EWAS). Light is coupled into one side of the microscope slide and collected on the opposite side, as shown in Figure 2, with the microscope slide acting as a planar waveguide. Traditional absorbance spectroscopy measured with illumination perpendicular to the slide surface has a short path length equal to the thickness of the applied layer and interacts with a small area of the slide surface (Figure 2). EWAS employs the width of the slide to obtain a 2.54 cm path length in addition to using a 1-cm wide fiber so that the measurement is acquired from a large area of the surface. The larger path length improves the signal-to-noise ratio while the large surface area involved eliminates the discrepancies between surfaces due to production in small quantities. EWAS has the additional advantage that the sample is not in the light path, so correction for sample absorbance is not necessary.

The experimental setup shown in Figure 3 uses a blue, light emitting diode as a light source (λ_{max} = 434 nm, FWHM = 83 nm, Kingbright, City of Industry, CA) coupled directly into the edge of the microscope slide. On the opposite side

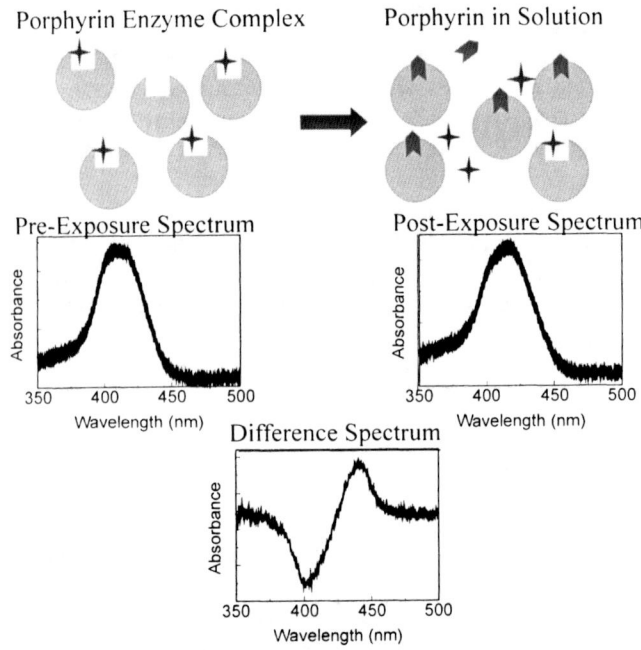

Figure 1. The interaction of analyte with the porphyrin-enzyme complex results in changes in the porphyrin absorbance spectrum.

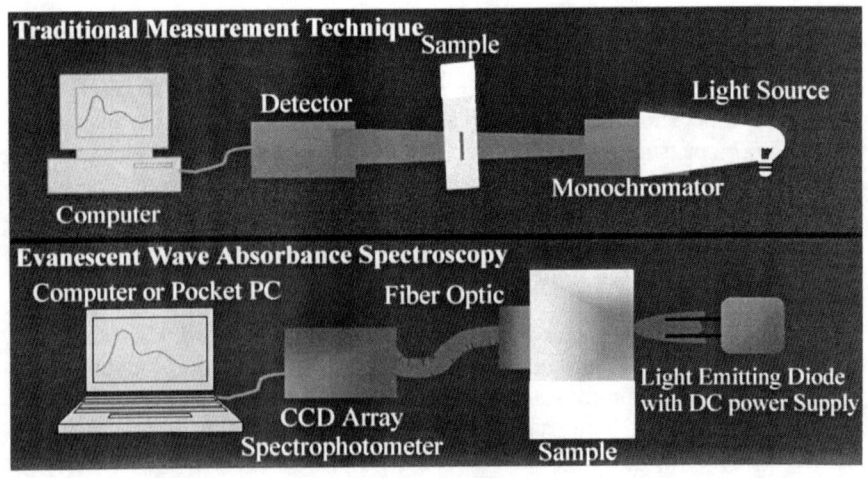

Figure 2. Top: Standard absorption measurement; Bottom: evanescent wave absorbance spectroscopy.

of the slide, a linear to circular fiber optic bundle (Dolan-Jenner, St. Lawrence, MA) collects the light and the transmission spectrum of the slide is measured using an Ocean Optics USB2000 CCD array spectrophotometer (Dunedin, FL). This measurement, together with a transmission spectrum for a clean microscope slide (reference) and a baseline measurement taken with no illumination (dark), can be used to calculate a pre-exposure absorbance spectrum for the sample slide.

Liquid samples are applied to the slide surface as 200 µl drops in 50 mM pH 7 sodium phosphate buffer (NaPi) with the exception of GB, which is in deionized water. Excess liquid is blotted away from the slide surface and a transmission spectrum collected. Vapor phase samples are applied by directing a gas stream across the slide surface. The difference spectrum shown in Figure 1 is used to facilitate analysis of the changes in the absorbance spectrum. It is calculated as the point-by-point subtraction of the pre-exposure absorbance spectrum from the post-exposure absorbance spectrum.

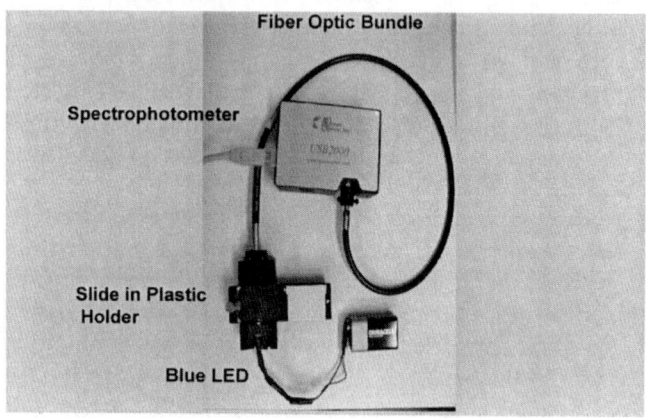

Figure 3. The experimental setup for EWAS.

The USB2000 can be used with a Pocket PC, desk top, or laptop computer. Data presented here were collected and analyzed (Grams/32, Galactic Industries, Salem, NH) on a laptop computer. Linear fittings were performed using PSI-Plot (v 7.0b).

Immobilization

Enzymes are an expensive component of a biosensor. Conserving enzyme and extending lifetime are two benefits arising from immobilization onto a support. Immobilization also increases the local enzyme concentration and maintains the proximity of the catalyzed reaction to the transducer. Covalent immobilization involving bond formation between a functional group on the surface of the protein and a glass support is often accomplished through the use of functionalized silane linkages, with amino-silane being a popular choice. Glutaraldehyde or 1,5-pentanediol has two chemically reactive terminal aldehyde groups. This molecule is commonly used as a biocide due to the cross-linking of proteins on cell surfaces that occurs when exposed to glutaraldehyde. When glutaraldehyde is applied at high concentrations to an amino-silane functionalized glass surface, it is possible to achieve a surface with many of these molecules bound so that one aldehyde group remains free to react with another amine. Exposure of this "activated" surface to a protein or an other amino-bearing molecule will result in attachment of that molecule to the glass surface via the five carbon chain of the glutaraldehyde molecule.

In many instances, the enzyme shows reduced activity when in close proximity to the slide surface. To address this problem, a molecule, bearing amino groups, can be used as a spacer. Possible candidates are arginine, poly-lysine, 1,12-diaminododecane, and similar molecules or amino-terminated dendrimers such as those of the PAMAM family which afford a greater density of amino residues. The protocol presented here is effective for the immobilization of glucose oxidase, organophosphorus hydrolase, acetyl-cholinesterase, and butyrylcholinesterase. An alternate protocol was developed for the immobilization of carbonic anhydrase.

Though covalent immobilization via glutaraldehyde-activated, amino-silane groups has been the method of choice for this work (12), it is not the best method for every enzyme or application. The immobilization protocol should consider the limitations and properties of the immobilization method and the support. Covalent immobilization is used here to increase enzyme stability and prevent diffusion issues in the resulting surface. One disadvantage of chemical immobilization is that is can distort enzymatic function. Other methods of immobilization include adsorption, encapsulation, cross-linking, and entrapment. Covalent immobilization can be accomplished via peptide bonds, diazo or isourea linkages, or alkylation reactions, to name a few. The amino acid residues involved in catalysis and substrate binding should remain unaffected upon immobilization, so the choice of bond type is influenced by the active site conformation for the enzyme under consideration.

Acetylcholinesterase (AChE, type V-S from electric eels), butyryl-cholinesterase (pseudocholinesterase from horse serum, E.C. 3.1.1.8), and glutaraldehyde were obtained from Sigma (St. Louis, MO). Organophosphorus hydrolase (EC 3.1.8.1; OPH, phosphotriesterase) was obtained as described previously (17). Amino-terminated Starburst® (PAMAM) dendrimer (generation 4) was obtained from Aldrich (Milwaukee, WI). ProbeOn™ Plus microscope slides were obtained from Fisher Biotech (Pittsburgh, PA).

For the immobilization protocol used, ProbeOn™ Plus surfaces were reacted with glutaraldehyde at 0.17 M in 50 mM pH 8 sodium phosphate buffer (NaPi) for 25 min followed by rinsing with PBS (50 mM sodium phosphate 0.1 M NaCl solution at pH 9). PAMAM dendrimer generation 4 (6.4 mM in 50 mM pH 8 NaPi) was then applied for 90 min, followed by rinsing with PBS and reacting with 1 M TRIS pH 9 for 20 min. The PAMAM, rinse, TRIS, rinse sequence was then repeated, followed by application of enzyme at 200 nM in 50 mM pH 8 NaPi for 90 min. Finally, the slides were reacted with 1 M TRIS pH 9 for 40 min and rinsed with 50 mM pH 7 NaPi.

Storage of the surfaces under mild vacuum greatly extended the viable lifetime of the immobilized enzymes. The surfaces prepared were stored at room temperature, after vacuum packing in three-layer, food saver bags using a FoodSaver (Vac360) from Tilia (San Francisco, CA). The enzymatic activity of the acetylcholinesterase surface was monitored using the Ellman assay, with more than 80% of the original enzymatic activity remaining after one year (*13*). Organophosphorus hydrolase surfaces maintained enzymatic activity for more than 250 d (*17*).

The porphyrins used as colorimetric indicators in these detection protocols (Frontier Scientific, Logan, UT) were added to the surface shortly before use by applying a solution to the slide for 10 min, then rinsing the unbound porphyrin away. Exact conditions varied for each of the porphyrins used. Long-term storage of the surfaces after binding the porphyrin resulted in degradation of the surface, likely due to porphyrin-catalyzed reactions. The acetylcholinesterase surfaces, when stored under vacuum after porphyrin incorporation, showed predicted detection capabilities for up to 55 d (*13*).

Organophosphorus Hydrolase Surfaces

OPH catalyzes the hydrolysis of organophosphonates with P-O, P-F, P-S, and P-CN phosphoryl bonds such as those of coumaphos, sarin, VX, and paraoxon. Hydrolysis is facilitated by two metal atoms in the active site and results in the release of two protons (*19*). In the native dimeric form of the enzyme, zinc is coordinated to histidine residues in the active site. Reconstitution of the apo-enzyme with cobalt or cadmium results in higher enzymatic activity than that observed in the zinc-metalloenzyme (*20*).

The enzymatic activity of OPH has been shown to be stable at temperatures up to 50 °C, with rates of hydrolysis exceeding those of chemical hydrolysis by more than 40-fold (*21, 22*), making OPH useful in decontamination. OPH has also been employed for the detection of OPs based on techniques including optical, acoustic, potentiometric, and amperometric for monitoring pH change or production of *p*-nitrophenol, chlorferon, or F⁻ (Table I).

Table I. Selected organophosphonates detected using cholinesterase and/or organophosphorus hydrolase based techniques

Analyte	Enzyme	References
Coumaphos	ChE	(23, 24)
	OPH	(17, 25, 26)
Diazinon	ChE	(27-29)
	OPH	(17, 25, 30, 31)
Dichlorvos	ChE	(29, 32–34)
	OPH	(25, 30)
Diisopropyl	ChE	(35)
fluorophosphate	OPH	(36)
DFP	OPH	(37)
Malathion	ChE	(33, 38)
	OPH	(17)
Methamidophos	ChE	(39)
Paraoxon	ChE	(33, 34, 38, 40)
	OPH	(17, 25, 26, 30, 31, 36, 41–43)
Parathion	ChE	(27, 28, 38, 40)
	OPH	(25, 26, 30, 31, 41–43)
Trichlorfon	ChE	(23, 24)

An OPH-based surface was developed for the direct detection of OPs using copper meso-tri(4-sulfonato phenyl) mono(4-carboxy phenyl) porphyrin (CuC_1TPP) as a colorimetric indicator of OP binding by the enzyme (17). CuC_1TPP (15 μM in 50 mM NaPi pH 7) is applied to an immobilized OPH surface and allowed to interact for 20 min before rinsing unbound porphyrin away (50 mM NaPi pH 7). Exposure of the surface to OPs is accomplished by applying 200 μl of the analyte at the desired concentration to the surface and blotting the excess solution away. A characteristic peak for the interaction of CuC_1TPP with immobilized OPH at 412 nm was observed in the absorbance spectrum of the surface. Though the interaction of CuC_1TPP with OPH is through mixed-type inhibition (17), exposure to diazinon, malathion, paraoxon, or coumaphos results a loss in absorbance at 412 nm. The change in absorbance at 412 nm shows log-linear dependence on OP concentration (slopes vary by analyte). Limits of detection for each analyte are shown in Table II.

This detection protocol is broad spectrum since any OP will result in a positive signal. In the case of analytes that yield a hydrolysis product with an absorption spectrum such as coumaphos (chlorferon, 348 nm) and paraoxon

(p-nitrophenol, 410 nm), identification may also be made on the basis of the appearance of new absorbance peaks indicative of the product.

Table II. Detection limits for various analytes *(12–15, 17, 18)*

Enzyme	Porphyrin	Analyte	LOD (ppb)
Acetylcholinesterase	monosulfonate tetraphenyl porphyrin	acetylcholine	NA[a]
		tetracaine	0.250
		eserine	0.037
		galanthamine	0.050
		scopolamine	0.100
		diazinon	0.045
		Triton X-100	83.000
		GB (liquid)	0.100
		GB (vapor)	0.25 ng[b]
Butyrylcholinesterase	monosulfonate tetraphenyl porphyrin	eserine	0.050
		amitriptyline	0.072
		drofenine	0.091
		Triton X-100	40.000
Organophosphorus hydrolase	copper meso-tri(4-sulfonato phenyl) mono(4-carboxy phenyl) porphyrin	paraoxon	0.007
		coumaphos	0.250
		diazinon	0.800
		malathion	1.000

[a] not available; potential for detection of acetylcholine indicated, but the limit of detection for this analyte is not reported. [b] quantity applied to sorbent.

Cholinesterase Surfaces

The enzymes acetylcholinesterase and butyrylcholinesterase are competitively inhibited by many organophosphonates, including compounds such as insecticides, fungicides, herbicides, and chemical warfare agents. Acetylcholinesterase is responsible for the termination of nerve impulse transmission at the cholinergic synapses. The enzyme-catalyzed reaction involves the hydrolysis of acetylcholine to acetate and choline. The rate of this reaction in the presence/absence of inhibitor is an indicator of inhibitor presence (see Table I). The reaction rate can be monitored through pH change, electron transfer, thiocholine production (when acetylthiocholine is used as the substrate), or O_2 consumption or H_2O_2 production (when choline oxidase is used in a subsequent reaction) employing optical, amperometric, or potentiometric methods (Table I). Detection of inhibitors, regardless of the technique used to measure the reaction rate, involves several minutes and is subject to fluctuations in enzymatic activity related to changes in conditions such as pH and temperature as well as to the vitality of the enzyme. Despite the potential problems, a large amount of literature is available on enzyme-based sensors of this type for the detection of OPs with limits of detection ranging from micromolar to picomolar levels (Table I).

Sensor systems based on the reversible, competitive inhibition of AChE and BChE by monosulfonate tetraphenyl porphyrin ($TPPS_1$) were developed. $TPPS_1$ (350 μM in 50 mM pH 7 NaPi with 25% ethanol) was applied to surfaces of immobilized enzymes and allowed to interact for 20 min before rinsing the excess solution off with excess buffer (50 mM NaPi pH 7). $TPPS_1$ was found to be a competitive inhibitor of cholinesterase activity for both AChE and BChE (*12, 14, 15*). The absorbance spectrum was unique for the AChE-$TPPS_1$ complex (TAC) as compared to that of the BChE-$TPPS_1$ complex (TBC) with characteristic peaks at 446 nm for TAC and 412 nm for TBC. When $TPPS_1$ was applied to a surface bearing both AChE and BChE (TABC), the characteristic peaks for both TAC and TBC were observed in the absorbance spectrum.

The response of each of the three surfaces to a range of inhibitors was investigated using cholinesterase-inhibiting drugs such as galanthamine and scopolamine (Table II) as model compounds. Log-linear dependence on concentration of the change in absorbance was observed at the characteristic peak for the TAC and TBC surfaces, with limits of detection as low as 0.037 ppb. The TABC surface showed a loss in absorbance at 446 nm when exposed to analytes inhibiting AChE but not BChE, at 421 nm when exposed to analytes inhibiting BChE but not AChE, and at both 421 nm and 446 nm when exposed to analytes inhibiting of both enzymes. The concentration dependence of the loss in absorbance at each enzyme-specific peak and the limit of detection for each compound agreed with values for enzymes immobilized alone (TAC and TBC surfaces). Exposure of the TAC to inhibitors of BChE only or to non-competitive inhibitors of AChE resulted in no change in absorbance. Similarly, exposure of the TBC to inhibitors of AChE only resulted in no change in absorbance.

The TAC surface response to the organophosphate diazinon was also investigated. Exposure of the TAC surface to diazinon produced the expected loss in absorbance at 446 nm with an LOD of 0.045 ppb (150 pM). The limit of detection for an analyte under a given set of conditions (temperature, pH, ionic strength of the solution) was found to be dependent on the IC_{50} for the analyte under similar conditions. The IC_{50} for an inhibitor is the inhibitor concentration at which 50% of enzymatic activity is inhibited.

Exposure of the TAC surface to GB in solution yielded a loss in absorbance at 446 nm. The limit of detection achieved was somewhat higher than expected at 0.1 ppb. Experiments had indicated that the LOD would be lower than that of diazinon. The discrepancy is likely due to the fact that the sample solutions were comprised of 17 MΩ water with traces of isopropyl alcohol, rather than the carefully buffered and pH-controlled solutions used for the other analytes. Exposure of the TAC surface to GB was accomplished by desorbing a known quantity of GB from a 3 mm diameter sorbent tube packed with Chromosorb 106 at 200 °C, using 50 mL/min breathable air for two min, while the effluent flowed across the slide surface. The change in absorbance of the TAC surface shows log-linear dependence on GB quantity spiked onto the sorbent tube, with a limit of detection of 250 pg GB.

The cholinesterase surfaces described here, as in the case of OPH, are broad spectrum. The TAC and TBC surfaces indicate only the presence of a competitive inhibitor of the enzyme and not its identity, so a positive result would be observed upon exposure to an organophosphate or carbamate pesticide, nerve agent, or drug with anti-cholinesterase activity. When the two enzymes are combined on a single surface, classes of compounds can be distinguished, that is, the surface discriminates between competitive inhibitors of AChE only, competitive inhibitors of BChE only, and competitive inhibitors of AChE and BChE. Inhibitor identification can be made by comparison of the change in absorbance at the AChE peak as compared to that for the BChE peak on the combined surface.

Discussion

The reversible inhibition of an enzyme by an association-sensitive colorimetric agent is the novel aspect of this detection technique. In detail, the porphyrin interacts with the enzyme in a manner that requires its dissociation to allow analyte binding. The porphyrin-enzyme complex absorbance spectrum is different from that of the porphyrin alone. Changes in the absorbance spectrum that occur upon dissociation of the porphyrin from the enzyme are used to indicate the presence of the analyte. In addition, the degree of change is analyte concentration dependent.

The use of a direct event for indication of analyte presence and concentration allow this detection technique to be used in real time. This, together with the broad nature of the detection protocols, makes the sensor system ideally suited for application as an early warning device. The surfaces can be used for repeated assays throughout their functional lifetimes and are exceptionally sensitive, with limits of detection for organophosphorus compounds below the current safe drinking water standards (*13, 14, 17, 18, 44*). In addition to detection of a variety of analytes in solution (*10, 12–18*), these enzyme based surfaces have been applied to the detection of GB vapor as well as gaseous carbon dioxide (*16, 18*).

Chemical weapons, such as tear gas, were first used in World War I (xylyl bromide, August, 1914). In 1925, The Geneva Protocol was the first treaty to prohibit the use of chemical weapons in warfare. In 1997, the United States ratified a treaty signed by 160 nations agreeing to abstain from the use of chemical weapons and to destroy all chemical weapons stockpiles within 10 years. Unfortunately, verifying compliance is impossible and it does not affect abstaining nations or terrorist factions. The potential for production of chemical weapons by small groups was clearly demonstrated by Aum Shinrikyo, reinforcing the need for detectors of chemical warfare agents for the protection of first responders and military personnel. The technology described here would be useful as a component of a sensor array with multiple detection protocols working in concert to facilitate identification of the threat agent and reduce potential false positive/negative readings. Currently available technologies for chemical detection range from chemical sensitive color-change paper to flame photometric detectors and gas chromatographic mass spectrometrometers (GC/MS). Many of these technologies are sensitive to the presence of interferents such as cologne, paint, and diesel fuel. The acetylcholinesterase surfaces described here do not give false positive readings in the presence of gasoline or diesel exhaust, cigarette smoke, or cologne. The Joint Chemical Agent Detector (JCAD) uses surface acoustic wave (SAW) technology to detect a range of nerve, blister, and blood agents. Detection of 100 $\mu g/m^3$ GB in 30 seconds has been demonstrated with the JCAD. This is the IDLH (immediately dangerous to life or health) level reported by CDC/NIOSH for GB. Other enzyme-based techniques have reported detection limits as low as 84 $\mu g/m^3$, with measurements requiring approximately 15 min (*45*). Detection of GB levels as low as 2.5 $\mu g/m^3$, as demonstrated with the acetylcholinesterase surfaces, could provide critical additional time to allow endangered personnel to take precautionary measures.

Literature Cited

1. Center for Disease Control and Prevention *Sarin (GB)* **2003**, www.bt.cdc.gov.
2. Reutter, S. *Environ. Health Persp.* **1999**, *107*, 985-990.
3. Bachmann, T. T; Leca, B.; Vilatte, F.; Marty, J. L.; Fournier, D.; Schmid, R. D. *Biosens. Bioelectron.* **2000**, *15*, 193-201.
4. Chen-Goodspeed, M.; Sogorb, M. A.; Wu, F. Y.; Raushel, F. M. *Biochemistry* **2001**, *40*, 1332-1339.
5. Harel, M.; Sussman, J. L.; Krejci, E.; Bon, S.; Chanal, P.; Massoulie, J.; Silman, I. *Proc. Natl. Acad. Sci. U.S.A.* **1992**, *89*, 10827-10831.
6. Sun, H.; El Yazal, J.; Lockridge, O.; Schopfer, L. M.; Brimijoin, S.; Pang, Y. P. *J. Biol. Chem.* **2001**, *276*, 9330-9336.

7. Mauzerall, D. *Biochemistry* **1965**, *4*, 1801-1810.
8. Shelnutt, J. A. *J. Phys. Chem.* **1983**, *87*, 605-616.
9. Schneider, H. J.; Wang, M. *J. Org. Chem.* **1994**, *59*, 7464-7472.
10. White, B. J.; Harmon, H. J. *Biochem. Biophys. Res. Commun.* **2002**, *296*, 1069-1071.
11. White, B. J.; Harmon, H. J. *Biosens. Bioelectron.* **2002**, *17*, 463-469.
12. White, B. J.; Legako, J. A.; Harmon, H. J. *Biosens. Bioelectron.* **2002**, *17*, 361-366.
13. White, B. J.; Legako, J. A.; Harmon, H. J. *Biosens. Bioelectron.* **2003**, *18*, 729-734.
14. White, B. J.; Legako, J. A.; Harmon, H. J. *Sens. Actuators B* **2003**, *91*, 138-142.
15. White, B. J.; Legako, J. A.; Harmon, H. J. *Sens. Actuators B* **2003**, *89*, 107-111.
16. White, B. J.; Legako, J. A.; Harmon, H. J. *Sensor Letters* **2004**, in Press.
17. White, B. J.; Harmon, H. J. *Biosens. Bioelectron.* **2005**, *20*, 1977-1983.
18. White, B. J.; Harmon, H. J. *Sensor Letters* **2005**, *3*, 1-5.
19. Raushel, F. M.; Holden, H. M. *Adv. Enzymol. R.A.M.B.* **2000**, *74*, 51.
20. Omburo, G. A.; Kuo, J. M.; Mullins, L. S.; Raushel, F. M. *J. Biol. Chem.* **1992**, *267*, 13278-13283.
21. Lejeune, K. E.; Dravis, B. C.; Yang, F. X.; Hetro, A. D.; Doctor, B. P.; Russell, A. J. In *Enzyme Engineering XIV;* Laskin, A. I.; Li, G.-X., Eds.; Annals of the New York Academy of Sciences; NYAS, New York, 1998, Vol. 864, pp. 153–170.
22. Munnecke, D. M. *Biotechnol. Bioeng.* **1979**, *21*, 2247-2261.
23. Ivanov, A. N.; Evtugyn, G. A.; Lukachova, L. V.; Karyakina, E. E.; Budnikov, H. C.; Kiseleva, S. G.; Orlov, A. V.; Karpacheva, G. P.; Karyakin, A. A. *IEEE Sensors J.* **2003**, *3*, 333-340.
24. Gogol, E. V.; Evtugyn, G. A.; Marty, J. L.; Budnikov, H. C.; Winter, V. G. *Talanta* **2000**, *53*, 379-389.
25. Rogers, K. R.; Wang, Y.; Mulchandani, A.; Mulchandani, P.; Chen, W. *Biotechnol. Prog.* **1999**, *15*, 517-521.
26. Mulchandani, A.; Pan, S. T.; Chen, W. *Biotechnol. Prog.* **1999**, *15*, 130-134.
27. Lee, H. S.; Kim, Y. A.; Cho, Y. A.; Lee, Y. T. *Chemosphere* **2002**, *46*, 571-576.
28. Kumaran, S.; Morita, M. *Talanta* **1995**, *42*, 649-655.
29. Wilkins, E.; Carter, M.; Voss, J.; Ivnitski, D. *Electrochem. Commun.* **2000**, *2*, 786-790.
30. Schoning, M. J.; Krause, R.; Block, K.; Musahmeh, M.; Mulchandani, A.; Wang, J. *Sens. Actuators B* **2003**, *95*, 291-296.
31. Mulchandani, P.; Mulchandani, A.; Kaneva, I.; Chen, W. *Biosens. Bioelectron.* **1999**, *14*, 77-85.

32. Evtyugin, G. A.; Ryapisova, L. V.; Stoikova, E. E.; Kashevarova, L. B.; Fridland, S. V.; Latypova, V. Z. *J. Anal. Chem.* **1998**, *53*, 869-875.
33. Hart, A. L.; Collier, W. A.; Janssen, D. *Biosens. Bioelectron.* **1997**, *12*, 645-654.
34. Andreescu, S.; Barthelmebs, L.; Marty, J. L. *Anal. Chim. Acta* **2002**, *464*, 171-180.
35. Makower, A.; Halamek, J.; Skladal, P.; Kernchen, F.; Scheller, F. W. *Biosens. Bioelectron.* **2003**, *18*, 1329-1337.
36. Simonian, A. L.; Grimsley, J. K.; Flounders, A. W.; Schoeniger, J. S.; Cheng, T. C.; DeFrank, J. J.; Wild, J. R. *Anal. Chim. Acta* **2001**, *442*, 15-23.
37. Mello, S. V.; Mabrouki, M.; Cao, X. H.; Leblanc, R. M.; Cheng, T. C.; DeFrank, J. J. *Biomacromolecules* **2003**, *4*, 968-973.
38. Campanella, L.; Achilli, M.; Sammartino, M. P.; Tomassetti, M. *Bioelectroch. Bioener.* **1991**, *26*, 237-249.
39. Lui, J.; Tan, M.; Liang, C.; Ying, K. B. *Anal. Chim. Acta* **1996**, *329*, 297-304.
40. Marty, J. L.; Mionetto, N.; Noguer, T.; Ortega, F.; Roux, C. *Biosens. Bioelectron.* **1993**, *8*, 273-280.
41. Richins, R. D.; Mulchandani, A.; Chen, W. *Biotechnol. Bioeng.* **2000**, *69*, 591-596.
42. Wang, J.; Chen, L.; Mulchandani, A., Mulchandani, P.; Chen, W. *Electroanalysis* **1999**, *11*, 866-869.
43. Mulchandani, A.; Mulchandani, P.; Chen, W.; Wang, J.; Chen, L. *Anal. Chem.* **1999**, *71*, 2246-2249.
44. National Research Council Committee on Toxicology *Guidelines for Chemical Warfare Agents in Military Field Drinking Water*; National Academies Press: Washington, DC, 1995; Vol. 10 pp 19-28.
45. Lee, W. E.; Thompson, H. G.; Hall, J. G.; Bader, D. E. *Biosens. Bioelectron.* **2000**, *14*, 795-804.

Chapter 5

Design of Sorbent Hydrogen Bond Acidic Polycarbosilanes for Chemical Sensor Applications

Eric J. Houser[1], Duane L. Simonson[1], Jennifer L. Stepnowski[2], Mike R. Papantonakis[1], Stuart K. Ross[3], Stanley V. Stepnowski[1], Eric S. Snow[1], Keith F. Perkins[1], Chet Bryant[4], Peter LaPuma[4], Gary Hook[4], and R. Andrew McGill[1]

[1]Naval Research Laboratory, 4555 Overlook Avenue, SW, Washington, DC 20375
[2]NOVA Research, Inc., 1900 Elkin Street, Suite 230, Alexandria, VA 22308
[3]Detection Department, Defense Science and Technology Laboratory, Porton Down, Salisbury SP4 0JQ, United Kingdom
[4]Uniformed Services University of the Health Sciences, 4301 Jones Bridge Road, Bethesda, MD 20814

>New hydrogen bond acidic polycarbosilanes have been developed for chemical sensing applications. The sorptive properties of the functionalized polycarbosilanes have been evaluated on a number of chemical sensing platforms including surface acoustic wave resonators, single wall carbon nanotube sensors and an Attenuated Total Reflectance Fourier Transform Infrared spectrometer. The hydrogen bond acidic nature of the polymers makes them highly sorbent toward basic vapors such as organophosphonates and polynitroaromatics.

Introduction

There currently exists a strong interest in developing rapid methods for detecting and identifying potential threat signature vapors such as organophosphonates, polynitroaromatics and a wide range of toxic industrial materials. As a result, the number of efforts to develop high performance chemical detection and analysis systems and related components continues to grow. There are many analytical chemistry techniques that utilize the interactions between analyte vapors and a chemically tailored stationary phase as a means of carrying out analyte separation, sensing or collection. Due to the relative ease of modifying their physicochemical properties, polymers continue to play a major role in the area of sorbent materials. The need for tailoring physicochemical properties becomes even more critical as one scales down the size of system components to micron or smaller dimensions since the relative proportion of material interfaces grows with decreasing device size. The use of nanoscale sensing devices, such as a single wall carbon nanotube, can place additional requirements on the nature of sorbent coatings since the materials in direct contact with the nanoscale sensing element can effect fundamental electronic or other fundamental property changes in the device. In general, for a polymer film to be optimal as a sorptive material it should have: (1) high thermal and chemical stability to provide device longevity, (2) a high density of chemically specific functional groups that impart sensitivity, selectivity and reversibility in vapor sorption, (3) physical properties that allow rapid vapor penetration and release through the film for fast response and recovery kinetics, and (4) the ability to form stable interfaces with other materials and surfaces present in a given device or system.

For analytical applications requiring the concentration or collection of basic vapors such as organophosphonates or nitroaromatics, hydrogen bond acidic polymers are a natural choice for the sorbent phase (*1–6*). However, hydrogen bond acidic coatings suitable for chemical sensor applications are not presently commercially available and must be prepared. Until recently, most hydrogen bond acidic polymers have been fluoroalcohol-substituted polyethers or polysiloxanes. Although these materials provide the desired functional groups and physical properties for chemical sensor applications, they are less than optimal due to the presence of hydrogen bond basic sites (oxygen atoms) present in the polymer backbone.

This work describes recent advances in the preparation of hydrogen bond acidic polycarbosilanes and their application as chemical sensor coatings. We have prepared hydrogen bond acidic polymers based on polycarbosilanes with the goal of improving upon the chemical and thermal properties of this class of functionalized polymer (*3–5*). Results pertaining to the preparation of selected model compounds are also described.

Experimental

Materials

All manipulations were carried out using standard Schlenk techniques. Tetrahydrofuran (THF) was distilled from sodium/benzophenone under argon. Chloroform was vacuum distilled from calcium hydride. Unless otherwise noted, chemicals were purchased from Aldrich Chemical Company and used as received. Hexafluoroacetone (HFA) was purchased from Oakwood Products or Aldrich Chemical Company and used as received. All other chemicals were reagent grade.

FTIR spectra were recorded on a Digilab FTS 7000 series FTIR spectrometer using NaCl plates. NMR spectra (300 MHz for ^{1}H, 75 MHz for ^{13}C with broadband decoupling) were collected with a Bruker ATS instrument. Size exclusion chromatograms (SEC) were recorded in THF using a Varian ProStar pump with a flow rate of 1.0 mL/min through a series of three Polymer Laboratories PLgel 10 μm MIXED-B columns (300 x 7.5 mm) with a Polymer Laboratories PL-ELS 1000 light scattering detector. SEC data were referenced to Polymer Laboratories Easical PS-1 polystyrene standards (MW 580–2,560,000).

Frequency responses of polymer coated surface acoustic wave (SAW) devices to vapor challenges were recorded using a Stanford Research Systems Model SR620 Universal Time Interval Counter. SAW devices used in this work were 312 MHz dual port resonators from RF Monolithics (Part No. RP1236). Polymer coatings were applied to the exposed SAW devices by spray coating a 0.1% by weight $CHCl_3$ solution of the polymer to a device frequency shift of 400–450 kHz.

Single wall carbon nanotube (SWNT) network sensors were prepared as previously described (7–9). The carbon nanotube network (CNN) sensors were spray coated with a 0.1% by weight solution of polymer to a thickness of approximately 100 nm. These CNN sensors can be monitored simultaneously in capacitive and resistive mode. The capacitance was measured by applying a 0.1 V, 30 kHz, AC voltage between a conducting Si substrate and a SWNT network deposited on a thermal SiO_2 layer. The induced AC current was measured using a Stanford Research Systems SR830 lock-in amplifier.

Synthesis of Monomers

Aryl substituted silacyclobutanes were prepared by reaction of 1,1-dichloro-1-silacyclobutane or 1-chloro-1-methyl-1-silacyclobutane with suitable Grignard species. Aryl- and alkenyl-substituted silanes were prepared by reaction of allyldichlorosilane with suitable Grignard reagents or by reaction of an organodichlorosilane with excess allylmagnesium bromide.

Synthesis of Parent Polycarbosilanes

Linear polycarbosilanes with pendant aryl groups were prepared by platinum-catalyzed or thermal ring opening polymerization of aryl-substituted silacyclobutanes. Linear polycarbosilanes with pendant aryl groups were also prepared by platinum-catalyzed hydrosilation reactions of diorganoallylsilanes. Hyperbranched polycarbosilanes with aryl and alkenyl pendant groups were prepared by platinum-catalyzed hydrosilation of AB2 organodiallylsilane monomers or by treatment of an intermediate poly(dichlorosilylene methylene) with excess aryl- or alkenyl-substituted Grignard reagents (*10–13*).

Reaction of Hexafluoroacetone (HFA) and Polycarbosilanes with Alkenyl Pendant Groups

A sample of parent polymer (1–2 g) was placed in a steel cylinder and the cylinder was evacuated. Chloroform (~100 mL) and HFA (3–5 g) were vacuum transferred into the cylinder sequentially. The mixture was heated to 50 °C, with stirring, for 24 h. The volatiles were removed in vacuo and the polymer extracted with $CHCl_3$, filtered through Celite and the solvent removed, leaving a fluoroalcohol-substituted polymer.

Reaction of HFA and Polycarbosilanes with Aryl Pendant Groups

A sample of parent polymer (1–2 g) was mixed with a catalytic amount of $AlCl_3$ (0.10 g), under inert atmosphere, and placed in a steel cylinder under vacuum. HFA (3–5 g) was vacuum transferred into the cylinder. The mixture was heated to 65 °C, with stirring, for 48 h. The volatiles were removed in vacuo and the polymer extracted with $CHCl_3$, filtered through Celite and the solvent removed leaving a fluoroalcohol-substituted polymer.

Synthesis of Fluoroalcohol-Substituted Naphthalene Model Compounds

A sample of 2-ethylnaphthalene was reacted with HFA in the presence of catalytic amounts of $AlCl_3$ at 60 °C for 48 h. Volatiles were removed and the products extracted with diethyl ether. Purification of the crude reaction mixture by column chromatography (SiO_2) with a gradient of acetone (0–30%) in hexanes provided the 6,8-, 4,6- and 4,6,8-1,1,1,3,3,3-hexafluoro-2-hydroxy-2-propyl substituted 2-ethylnaphthalenes in 75, 20 and 3% yields, respectively.

Results and Discussion

Polycarbosilanes with linear, hyperbranched or dendritic structures can be prepared by hydrosilation, reductive coupling or ring opening polymerization routes (*10–15*). Hydrosilation routes to hyperbranched polycarbosilanes with a wide range of structural variation and pendant groups are well established (*14, 15*). The variety of available structural motifs and the flexibility in choosing pendant group identity makes polycarbosilanes highly useful as a foundation for the preparation of a variety of functionalized sorbent materials with desirable physicochemical properties. For the preparation of hydrogen bond acidic polycarbosilanes, it is often desirable to select at least some of the pendant groups to be terminal alkenyl, aryl or alkylaryl groups to provide sites within the polymer for subsequent reaction with HFA to form pendant fluoroalcohol species (*16–18*). Representative reaction pathways to linear and hyperbranched polycarbosilanes with alkenyl, aryl or alkylaryl pendant groups are shown in Figure 1.

Silacyclobutane monomers with aryl and alkylaryl substituents were prepared in multi-gram quantities from reaction of commercially available 1,1-dichloro-1-silacyclobutane and 1-chloro-1-methyl-1-silacyclobutane precursors with suitable Grignard reagents. Allyl-substituted AB and AB_2 type monomers were prepared from treatment of intermediate diorganochlorosilanes and organodichlorosilanes, respectively, with allylmagnesium bromide. Selected structures of the monomers used in this work are represented in Figure 2.

The platinum-catalyzed or thermal ring opening polymerization of the substituted silacyclobutane monomers yields linear polycarbosilanes. The platinum-catalyzed hydrosilations of the AB (allyl) and AB_2 (diallyl) monomers yield linear and hyperbranched polycarbosilanes, respectively. A wide range of random copolymers are readily available from polymerization of mixtures of the silacyclobutane or AB and AB_2 monomers. The preparation of random copolymers is often of interest for achieving desired physical properties.

A convenient synthetic pathway for the preparation of fluoroalcohol substituted polycarbosilanes is by reaction of suitable parent polymers with HFA. The reaction of HFA with unsaturated organic species, such as terminal alkenes or arenes, to give fluoroalcohol compounds is well known (*16–18*). The reaction between polymers with terminal alkene pendant groups and HFA proceeds readily in $CHCl_3$. In contrast, the reaction between HFA and polymers with pendant arene groups often requires catalytic amounts of a Lewis acid, such as $AlCl_3$, in order to achieve higher levels of ring substitution. Through sensor evaluations, we have found that polymers with pendant terminal alkenes provide higher sorption materials for organophosphonates, while polymers with naphthalene pendant groups tend to be best for nitroaromatic sorption. The resulting fluoroalcohol-substituted polycarbosilanes are highly viscous oils or solids that are soluble in common polar organic solvents. Selected structures of the functionalized polymers described herein are represented in Figure 3.

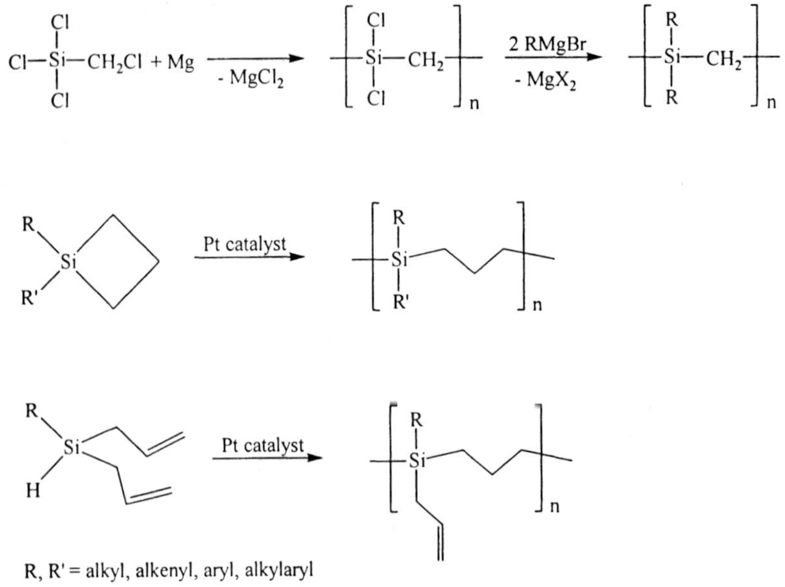

R, R' = alkyl, alkenyl, aryl, alkylaryl

Figure 1. Selected synthetic routes to polycarbosilanes.

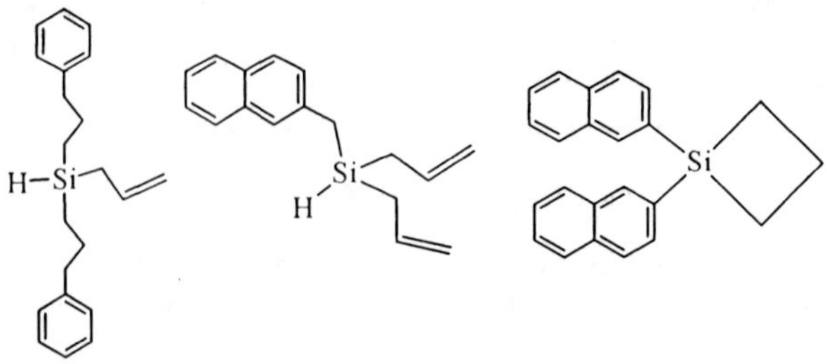

Figure 2. Structures of representative monomers used in the preparation of parent polycarbosilanes.

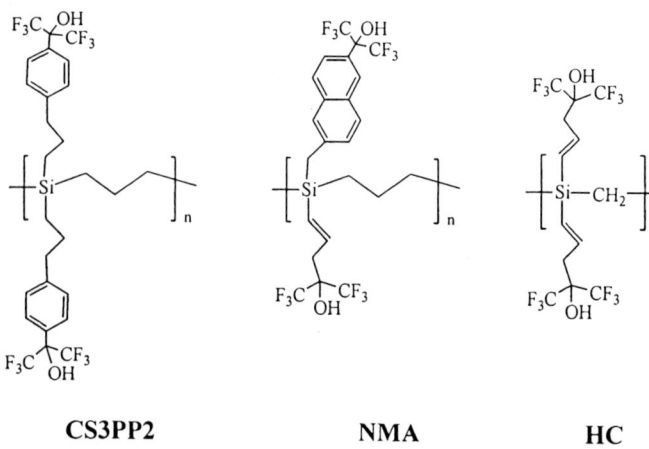

Figure 3. Structures of selected hydrogen bond acidic polycarbosilanes.

The presence of the fluoroalcohol groups in the functionalized polymer is readily verified by broad vO-H peaks in the FTIR spectrum in the 3300–3500 cm^{-1} region and C-F stretches near 1250 cm^{-1} and is shown in Figure 4. Although the reaction of HFA with terminal alkenes and phenyl species is a convenient method for the preparation of 1,1,1,3,3,3-hexafluoro-2-hydroxy-2-propyl pendant groups, this method is not general for other perfluorinated ketones. Thus, for the incorporation of other fluoroalcohol pendant groups or for functionalization of pendant groups other than arenes and terminal alkenes, alternative synthetic methodologies must be employed.

The thermal stability of HC has been preliminarily examined through repeated thermal cycling of polymer-coated MEMS structures between room temperature and 200 °C. Cycling of HC through this temperature range over 30,000 cycles has not resulted in any measurable performance degradation of the polymer sorptive properties. These observations are supported by thermogravimetric studies, which showed that HC begins thermal decomposition in air at approximately 230 °C.

In order to better understand the substitution patterns of the reaction of naphthalene-substituted polycarbosilanes with HFA, we examined the reaction of the model compound 2-ethylnaphthalene with HFA under similar reaction conditions as for the functionalization of the naphthalene-substituted parent polycarbosilanes. The reaction of 2-ethylnaphthalene with excess HFA in the presence of catalytic amounts of AlCl$_3$ resulted in the formation of 6,8-bis(1,1,1,3,3,3-hexafluoro-2-hydroxy-2-propyl)-2-ethylnaphthalene (major), 4,6-bis(1,1,1,3,3,3-hexafluoro-2-hydroxy-2-propyl)-2-ethylnaphthalene (minor), 4,6,8-tris(1,1,1,3,3,3-hexafluoro-2-hydroxy-2-propyl)-2-ethylnaphthalene (trace)

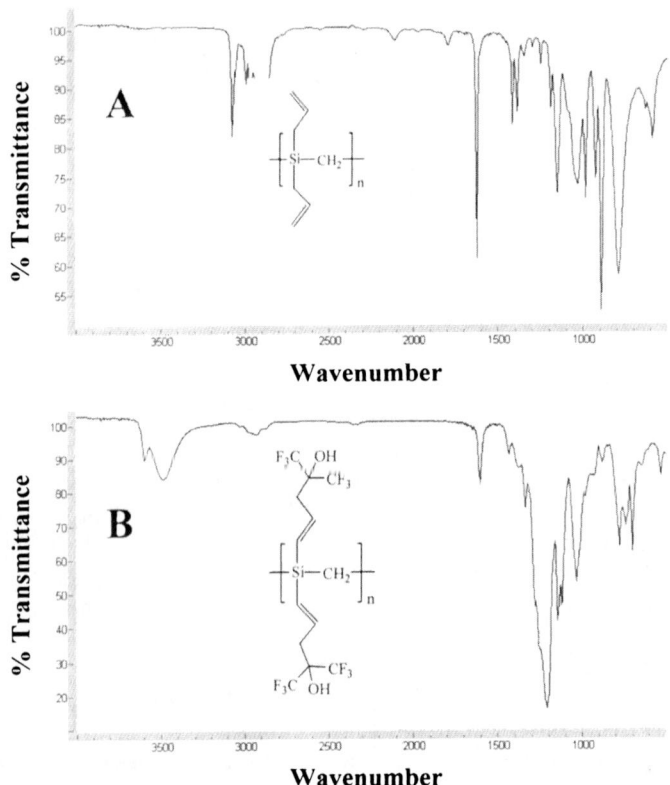

Figure 4. FTIR spectra of poly(diallylsilylene methylene) before (a) and after (b) reaction with HFA.

and 4,8-bis(1,1,1,3,3,3-hexafluoro-2-hydroxy-2-propyl)-2,6-diethylnaphthalene (minor). The structures of the 6,8-, 4,6- and 4,6,8-substituted isomers of 2-ethylnaphthalenes are represented in Figure 5. The presence of the substituted diethylnaphthalene product is likely due to a redistribution reaction of 2-ethylnaphthalene with $AlCl_3$ (*19*). The use of naphthalenes with bulkier substituents, such as 2-(trimethylsilylmethyl)-naphthalene, provided 6-(1,1,1,3,3,3-hexafluoro-2-hydroxy-2-propyl)-2-(trimethylsilylmethyl)-naphthalene as the major isolated product (*20*). These results indicate that the substitution of the 2-naphthyl pendant groups likely occurs first at the 6-position followed by further substitution at either the 4- or 8-positions of the naphthyl rings, with the 6,8-disubstitution being most likely. This pattern of substitution can be rationalized from known electrophilic substitution reactions of naphthalenes (*21, 22*). The ^{19}F NMR spectra of the hydrogen bond acidic naphthyl-substituted polycarbosilanes indicate that the pendant naphthyl rings are primarily monosubstituted, with 10–20% disubstitution occurring under the reaction conditions studied.

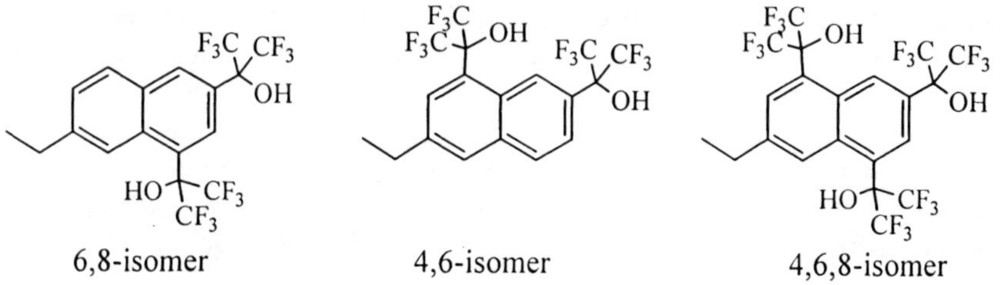

6,8-isomer   4,6-isomer   4,6,8-isomer

Figure 5. Structures of the primary substitution products from the reaction of 2-ethylnaphthalene with HFA.

Sensor Measurements

The sorptive properties of the hydrogen bond acidic polycarbosilanes were evaluated by coating the functionalized polymers onto a number of chemical sensor platforms, including SAW, SWNT-CNN sensors and a commercial Attenuated Total Reflectance Fourier Transform Infrared (ATR-FTIR) system. This range of sensor platforms demonstrates the general utility of chemoselective polymers in chemical sensing as SAW sensors primarily measure mass changes in the polymer coating during vapor exposure, CNN devices typically measure changes in the electronic properties of the nanotubes during vapor exposure and infrared spectral sensors detect analyte by the presence of specific spectral features in the infrared difference spectrum by comparison to stored spectral data libraries.

Simulant vapors were used for initial evaluation of the polymer sorptive properties. For detection of organophosphonates, dimethyl methylphosphonate

(DMMP), dimethyl phosphonate (DMHP) or diisopropyl methylphosphonate (DIMP) are commonly used as simulants. For nitroaromatic detection, 2,4-dinitrotoluene (2,4-DNT) is a commonly used simulant, although many TNT-based explosives often have a significant component of 2,4-DNT in the vapor signature.

Surface Acoustic Wave (SAW) Devices

A sorbent polymer film is applied to a SAW device by spray coating a dilute polymer solution onto the surface of the resonator while monitoring the frequency shift to estimate the thickness of the applied polymer film. The polymer-coated SAWs are then placed in a flow chamber and exposed to organic vapors under controlled conditions. These tests have shown that the new hydrogen bond acidic polycarbosilanes have high sensitivity to hydrogen bond basic vapors such as DMMP and 2,4-DNT. A comparison of the DMMP and 2,4-DNT vapor sorptive properties of HC and NMA revealed that these two polymers have similar responses to DMMP, while NMA is substantially more responsive to 2,4-DNT vapor (Figure 6). The observed decrease in SAW frequency is a result of the increase in mass of the polymer film associated with uptake of analyte vapors.

Comparative chemical sensor vapor testing of the hydrogen bond acidic polycarbosilanes with polysiloxane analogues has shown a significant enhancement in sensitivity toward hydrogen bond basic vapors relative to the

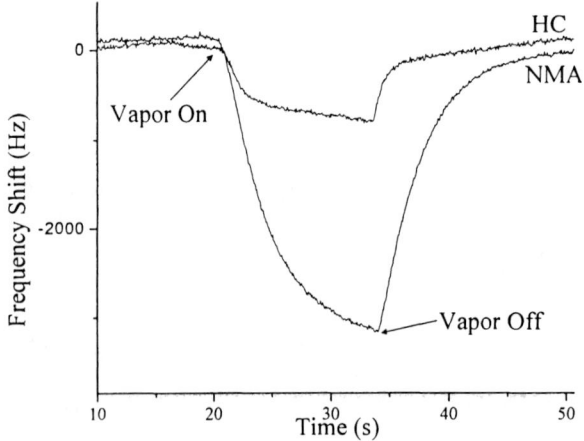

Figure 6. Parallel exposure of HC and NMA coated 312 MHz SAW devices to 50 ppb (100 mL/min) of 2,4-DNT vapor at 35 °C.

polysiloxanes. This enhancement in sensitivity is tentatively attributed to a lack of basic sites in the polycarbosilane backbone. While some evidence of stability problems has been reported for polysiloxanes with pendant phenol groups (6), the backbone of polycarbosilanes appears stable against reaction with fluoroalcohols and phenols under ambient conditions.

Single Wall Carbon Nanotube Network Sensors

Single wall carbon nanotube network sensors were prepared as described in the literature (8, 9). The carbon nanotube network can be monitored resistively and capacitively simultaneously or used as a chemically sensitive field effect transistor (chemFET). Although the electronic properties of the carbon nanotube are highly sensitive to the chemical environment in the immediate vicinity of the nanotube, the nanotube surface is relatively chemically inert and significant improvements in sensitivity can be obtained through application of polymer films to the surface of the carbon nanotube network devices. While direct covalent modification of the carbon nanotube surface is possible, this approach may be less desirable for carbon nanotube sensors as it results in substantial decreases in conductivity through nanotubes. It should also be noted that the choice of polymer film for coating single wall carbon nanotube sensors should be chosen with care, as application of the film can effect changes in the electronic properties of the nanotube network in addition to modifying the sorptive properties of the sensor. A comparison of changes in capacitance for HC-coated and self-assembled monolayer (SAM) treated single wall carbon nanotube devices during exposures to DMMP and DMHP is shown if Figure 7.

The sensor in the two upper traces of Figure 7 was coated with a SAM formed from silanization with trichloroallylsilane, followed by treatment with HFA. The sensor in the lower two traces was coated with ~ 100 nm-thick layer of the polymer HC. The amplitude and temporal responses are both useful in discriminating chemically similar analytes. While the HC polymer coated device is substantially more sensitive than the SAM coated device, the kinetics of the response are considerably slower for the polymer film, presumably due to diffusion of the analyte into the 100 nm thick film. Polymer film thicknesses on the order of 1–3 nm provide optimal vapor sorption kinetics for single wall carbon nanotube chemical sensors, although coating thicknesses above this range are still useful for extending the dynamic range of these devices.

ATR-FTIR Vapor Detection

The hydrogen bond acidic polycarbosilane HC was examined as a sorbent coating for chemical vapor detection by ATR-FTIR. For optimal sensitivity, ATR-FTIR instruments must have physical contact between a sufficient mass of analyte and the reflection interface of the instrument. This contact is easily achieved for liquid and solid samples, but it typically prohibits ATR-FTIR instruments from analyzing air samples directly. Using a thin layer of a hydrogen bond acidic polycarbosilane on the reflection surface, an ATR-FTIR instrument was able to detect and correctly identify the nerve agent simulants DIMP and DMMP with limits of detection (LOD) of 50 parts per billion (ppb) and 250 ppb, respectively, as described below.

The instrument used in this study was the TravelIR^{TM} (Smiths Detection), which employs a miniaturized Michelson interferometer and an integrated diamond ATR sample interface represented in Figure 8. An evanescent wave extends beyond the diamond surface and is partially attenuated by substances within 0.5–3.1 μm of the diamond surface in the range of 4000–650 wavenumbers. An embedded, on-board computer uses an automated search algorithm that compares infrared spectral features to digital spectral databases.

To test the instrument's ability to detect and correctly identify organophosphonate vapors, a layer of HC polymer was placed on the diamond surface. To deposit a thin layer of HC polymer on the diamond reflection surface, a 1 μL drop of 1% mixture of HC in chloroform was placed on the surface and allowed to evaporate. Figure 9 shows the spectral comparison of the simulants DIMP and DMMP to that of the HC polymer (bottom spectra). Once the diamond was coated with a film of HC polymer, a new background reading was taken to provide a level baseline before analyte sampling began. The HC film was exposed to known concentrations of DIMP or DMMP vapors for 8 min at flow rates of 1 LPM at 25 °C, which is shown in Figure 10.

The main spectral features that were identified for DIMP and DMMP were at 1000 and 1050 wavenumbers, respectively, assigned to the P=O stretches. Other peaks such as the asymmetric/symmetric methyl C-H stretches at 2975 and 2950 cm^{-1} could also be used to aid in the identification. The negative peaks between 1200–1300 wavenumbers correspond to the dominant C-F stretch from the HC polymer and are present as a result of the computer's overcompensation for these peaks when a new background was taken after the polymer was applied to the diamond window.

Requirements for the establishment of the LODs from this study were that the instrument could produce a spectrum with peaks that were at least 3 times greater than the surrounding noise, that could be visually analyzed and attributed to the stimulant being sampled, and that the instrument could correctly match to the onboard library with a statistically significant difference from the next closest match. In this case, the library matching was the limiting factor, while the instrument produced spectra that could be visually matched to the simulants at even lower levels.

Most of the principles associated with air sampling using SPME fiber sampling also applied to this sampling method. The thickness of the coating affected the sample time required for the polymer and simulant vapor to come into equilibrium, and thicker polymer layers resulted in more simulant vapor

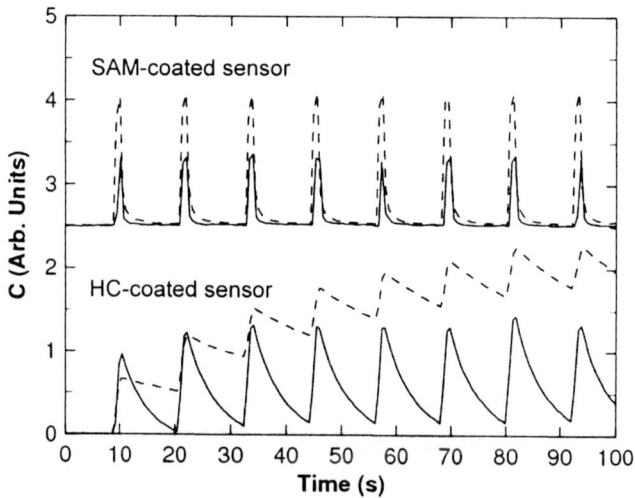

Figure 7. Temporal response of two SWNT chemicapacitors to repeated 2 s doses of two nerve agent simulants $(CH_3O)_2P(O)H$ (solid) and $(CH_3O)_2P(O)CH_3$ (dashed).

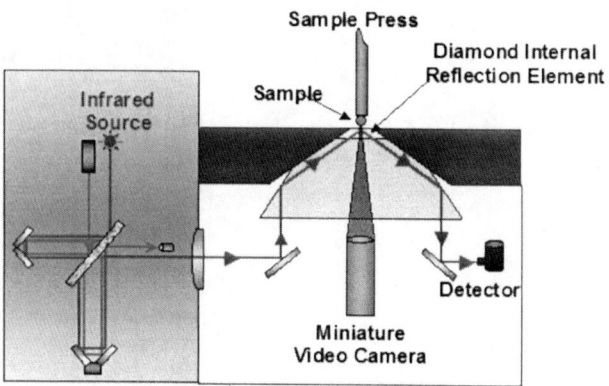

Figure 8. Schematic of TravelIR^{TM} Miniature ATR-FTIR Spectrometer.

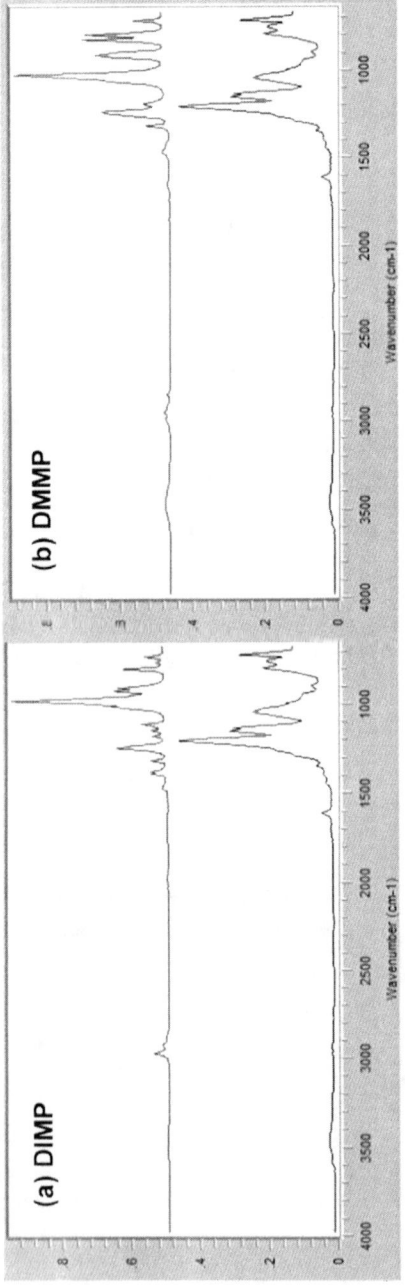

Figure 9. FTIR spectra of DIMP (a, top) and DMMP (b, top) compared to HC polymer (bottom).

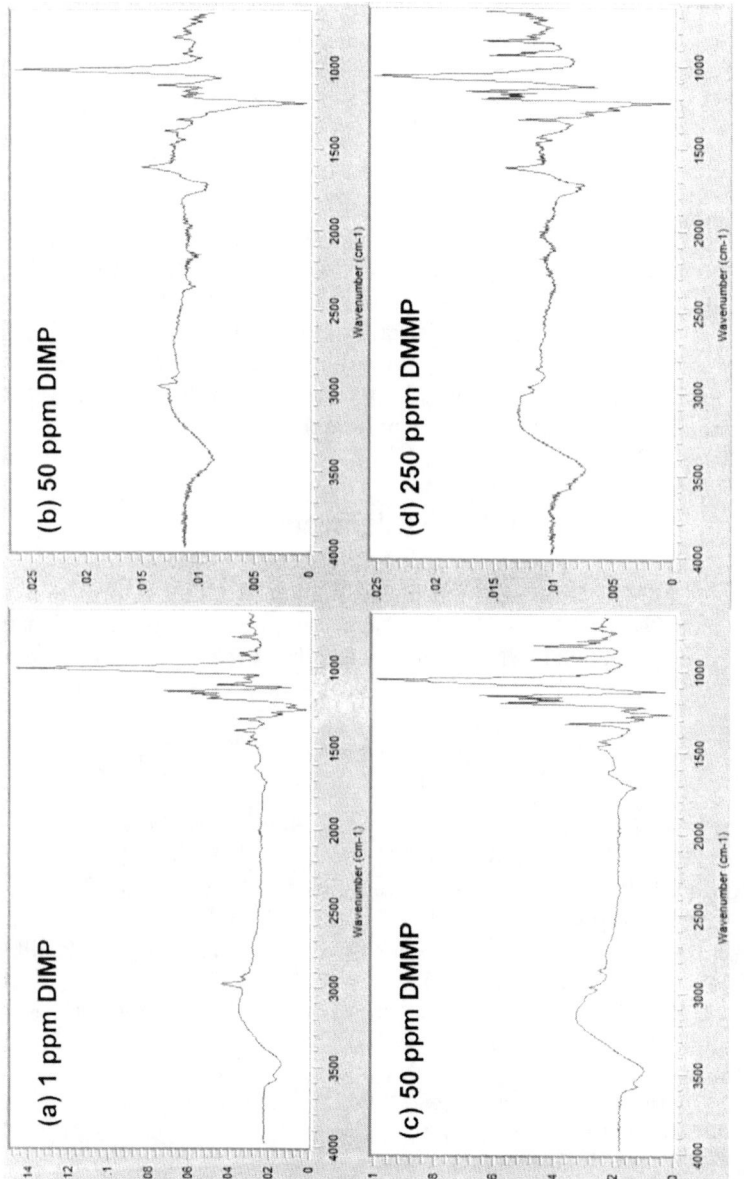

Figure 10. FTIR difference spectra of DIMP (a,b) and DMMP (c,d) sorbed by HC polymer.

collection (higher IR absorption by the sample). Sample time and flow rate also affected the sampling process and both factors could be increased to further lower the LOD. While the system as tested was not optimized for minimum limits of detection, this effort clearly demonstrated the ability to extend the capabilities of portable commercial ATR-FTIR instrumentation to gas phase detection.

Conclusions

Hydrogen bond acidic polycarbosilanes have been found to be promising materials for the concentration of organophosphonate and polynitroaromatic vapors in a number of sensor platforms. Fluoroalcohol-substituted polycarbosilanes show a higher sensitivity toward hydrogen bond basic vapors than related polysiloxane analogues, which is attributed to the absence of the hydrogen bond basic oxygen sites in the polycarbosilane backbone.

Acknowledgments

We gratefully acknowledge financial support for this effort from the Transportation Security Administration, Bureau of Alcohol, Tobacco, Firearms and Explosives, United States Marine Corps and the Office of Naval Research.

References

1. Grate, J. W.; Patrash, S. J.; Kaganove, S. N. *Anal. Chem.* **1999**, *71*, 1033.
2. McGill, R. A.; Abraham, M. H.; Grate, J. W. *CHEMTECH* **1994**, *24(9)*, 27.
3. Simonson, D. L.; Houser, E. J.; Stepnowski, J. L.; Nguyen, V.; McGill, R. A. *PMSE Preprints* **2003**, *88*, 546-547.
4. Simonson, D. L.; Houser, E. J.; Stepnowski, J. L.; Nguyen, V.; McGill, R. A. *PMSE Preprints* **2003**, *89*, 866.
5. Houser, E. J.; Mlsna, T. E.; Nguyen, V. K.; Chung, R.; Mowery, R. L.; McGill, R. A. *Talanta* **2001**, *54*, 469.
6. Hartmann-Thompson, C.; Hu, J.; Kaganove, S. N.; Keinath, S. E.; Keeley, D. L.; Dvornic, P. R. *Chem. Mater.* **2004**, *16*, 5357.
7. Kong, J.; Franklin, N. R.; Zhou, C.; Chapline, M. G.; Peng, S.; Cho, K.; Dai, H. *Science* **2000**, *87*, 622.
8. Novak, J. P.; Snow, E. S.; Houser, E. J.; Park, D.; Stepnowski, J. L.; McGill, R. A. *Applied Physics Letters* **2003**, *83(19)*, 4026.
9. Snow, E. S.; Perkins, F. K.; Houser, E. J.; Badescu, S. C.; Reinecke, T. L. *Science* **2005**, *307*, 1942.

10. Interrante, L. V.; Shen, Q. In *Silicon-Containing Polymers*, Jones, R. G.; Ando, W.; Chojnowski, J., Eds.; Kluwer Academic Publishers: Dordrecht, Netherlands, 2000.
11. Whitmarsh, C. K.; Interrante, L. V. *Organometallics* **1991**, *10*, 1336.
12. Rushkin, I. L.; Shen, Q.; Lehman, S. E.; Interrante, L. V. *Macromolecules* **1997**, *30*, 3141.
13. Interrante, L. V.; Rushkin, I.; Shen, Q. *Appl. Organomet. Chem.* **1998**, *12*, 695.
14. Yoon, K.; Son, D. Y. *Macromolecules* **1999**, *32*, 5210.
15. Lach, C.; Frey, H. *Macromolecules* **1998**, *31*, 2381.
16. Nagai, T.; Kumadaki, I. *J. Syn. Org. Chem. Jpn.* **1991**, *49(7)*, 624.
17. Urry, W. H.; Niu, J. H. Y.; Lundsted, L. G. *J. Org. Chem.* **1968**, *33*, 2302.
18. Farah, B. S.; Gilbert, E. E.; Sibilia, J. P. *J. Org. Chem.* **1965**, *30*, 998.
19. Olah, G.; Olah, J. A. *J. Am. Chem. Soc.* **1976**, *98*, 1839.
20. Unpublished results, Simonson, D. L.; Houser, E. J.; McGill, R. A.
21. Field, L. D.; Sternhell, S.; Wilton, H. V. *J. Chem. Ed.* **1999**, *76*, 1246.
22. Zhao, X. S.; Lu, M. G. Q.; Song, C. *J. Mol. Catal. A* **2003**, *191*, 67.

Chapter 6

Non-Aqueous Polymer Gels with Broad Temperature Performance

Joseph L. Lenhart[1,*], Phillip J. Cole[1,*], Burcu Unal[2], and Ronald C. Hedden[2]

[1]Sandia National Laboratories, Albuquerque, NM 87185
[2]Department of Materials Science and Engineering, Pennsylvania State University, University Park, PA 16802
[*]Corresponding authors: jllenha@sandia.gov; pjcole@sandia.gov

A gel is a physically or chemically cross-linked polymer that is highly swollen with solvent. While significant work has focused on aqueous hydrogels for biotechnology applications, hydrogels suffer from a limited operating temperature range due to the moderate freezing point and high volatility of water. In this work, a non-aqueous, chemically cross-linked polybutadiene gel has been designed that exhibits stable properties over a temperature range of -70 to 70 °C. A combination of swelling experiments, rheology, neutron scattering, and tack adhesion testing was utilized to characterize the gel properties over a broad range of temperatures. The methodology utilized to design the polybutadiene gel can be generalized to different gel materials and applications.

© 2008 American Chemical Society

Introduction

A polymer gel is a physically or chemically cross-linked polymer that is highly swollen by solvent. Mechanically the solvent creates a soft solid, which is easily deformable, yet still recovers from the deformation due to the elastic nature of the cross-links in the polymer (*1*). Polymer gels offer potential for a wide array of applications because the gel properties can be tuned by varying the polymer type, solvent type, and solvent loading. In addition, small molecule additives and fillers can be incorporated into the gel formulation to enhance the properties further. Polymer gels have been studied for many years (*2–5*), and early work focused on potential electro-active applications where mechanical motion was induced in the gel through an applied electric field (*6–8*). Recently, gels are emerging for consideration in a range of practical applications including biomedical technology (*9–12*), food and cosmetics (*13*), separations (*14*), magneto-rheological devices (*15, 16*), micro-valves (*17, 18*), electronic devices (*19–21*), robotics (*22*), catalysts (*23*), optical devices (*24–27*), sensors (*28–30*), oil and gas recovery (*31*), *etc*. A classic gel system is based on poly-(N-isopropylacrylamide) (PNIPAM), which is a thermo-responsive hydrogel that exhibits a volume-shrinkage transition near 33 °C (*32*). Due to the near-room-temperature responsiveness of PNIPAM hydrogels, they have been exploited in the biological community for many applications including controlled release (*33*), cellular adhesion (*34*), and nucleic acid purification (*35*). While hydrogels are useful for the biological and medical community, and the thermo-responsiveness of these gels can be exploited for certain devices, their use is limited for alternative applications such as micro-devices, homeland security sensors, electronics encapsulation, energy storage devices, information protection, protective coatings, controlled lubrication layers, etc., due to a narrow range of operating temperatures. In general, a need exists for polymer gels that maintain properties and performance over a broad range of temperatures. In this paper, we describe a polymer gel that performs over a temperature range of -70 to 70 °C, and a methodology to rapidly design a broad temperature performing gel that can be generalized to most applications.

The major material requirements for developing a gel that maintains performance over a broad range of temperatures are: (1) a flexible polymer backbone with a low glass transition temperature (T_g) and no potential to crystallize in the operating range of interest, (2) a low volatility solvent with a boiling point well above the desired highest operating temperature, (3) a solvent with no transitions over the operating range of interest, including glass or melting transitions, (4) a polymer-solvent system that is miscible over the entire operating range, (5) a gel that exhibits good adhesion over the operating temperatures to an array of different substrates, often without the ability enhance the adhesion through surface preparation techniques, and (6) an adjustable cross-linking reaction rate for different processes or applications. This paper will outline a non-aqueous polybutadiene gel that exhibits the above criteria.

Experimental

Two polybutadiene monomers were used to form the cross-linked network, one containing maleic anhydride functional groups (MA10 or MA5) and the other containing hydroxyl groups (R45). The hydroxyl groups react with the maleic anhydride groups to form an ester. The cross-linking reaction is performed in the presence of solvent, with solvent quantities varying from 0 to 60 mass percent. The major solvents utilized were dibutyl-phthalate (DBP) and bis(ethylhexyl)sebacate (BEHS). A catalyst, didecyl methylamine (DDA), promoted the cross-linking reaction. The gels utilized in this study were stoichiometric formulations, where the number of maleic anhydride groups equaled the number of hydroxyl groups. The gels were cured in sealed molds at 75 °C for 6 days. Rheological measurements showed that 6 days at 75 °C resulted in a completely cured gel material. The cured gels had an extractable content equal to the solvent loading plus 2 to 3 mass percent of unreacted polymer precursors. Fourier transform infrared spectroscopy verified the complete cure. The maleic anhydride and hydroxyl-functionalized polybutadienes were used, as received, from Sartomer Company. The solvents were used, as received, from Aldrich.

Rheological measurements were made with rectangular gel samples in a torsion geometry. The gel samples had dimensions of approximately 12 x 4.5 x 28 mm. The measurements were made on a Rheometric Scientific ARES instrument at a frequency of 1 Hz and a scan rate of 2 °C/min. An environmentally controlled chamber permitted determination of the modulus over temperatures from -100 to 70 °C. Strain sweeps were conducted at various temperatures to ensure that the modulus was independent of strain.

Scattering was performed at the National Institute of Standards and Technology Center for Neutron Research on the NG3 30 meter Small Angle Neutron Scattering (SANS) beam line for the DBP-based gels and on the BT5 (Ultra Small Angle Neutron Scattering) USANS beam line, which probes q values as low as $3 \cdot 10^{-5}$ Å^{-1}, for the BEHS-based gels. Deuterium labeled DBP and BEHS were purchased from CDN isotopes.

Tack adhesion measurements were made by curing the polybutadiene gel in a thin film (~ 0.5 mm thick) on a plate. A stainless steel probe (8 mm diameter) was brought into contact with the gel film and held for 60 s at a force of 500 g. The probe was pulled away from the gel film at a rate of 0.002 mm/s, while measuring the force-displacement curve. The temperature for the tack measurement was held constant utilizing an environmental chamber. The sample was allowed to equilibrate at the measurement temperature for 10 to 15 min prior to each measurement. The stress (force/probe area)/strain (displacement/film thickness) curve was integrated, multiplied by the gel sample thickness, and given a Poisson's ratio correction to get an effective tack adhesion energy.

Results and Discussion

Figure 1 shows the primary constituents of the gel formulation. To achieve the first criterion (flexible polymer), polybutadiene was chosen as the gel backbone. The gel is composed of two monomers. One is a polybutadiene backbone with maleic anhydride groups grafted to the backbone (MA5, $M_n \sim 5300$ g mol^{-1}, ~ 2.5 maleic anhydride groups per polymer chain, PDI ~ 2.5) (36). The other is a polybutadiene backbone with hydroxyl functional groups (R45, $M_n \sim 2800$ g mol^{-1}, ~ 2.5 hydroxyl groups per polymer chain, PDI ~ 2.5) (36). Polybutadiene is a flexible polymer chain with a T_g near -85 °C (37). The low functionality of the monomers ensures a loosely cross-linked gel with a low glass transition temperature. The maleic anhydride functionality on MA5 reacts with the hydroxyl functionality on R45 to form an ester linkage. To meet criterion six above (adjustable reaction rate), this reaction is promoted by a small fraction, < 2%, of a tertiary amine catalyst, DDA. The two solvents utilized in this study were DBP and BEHS, also shown in Figure 1. These solvents meet the second (low volatility) and third (no solvent transitions) criteria above, as both DBP and BEHS have high boiling points (340 °C at 760 mm Hg and 240 °C at 4 mm Hg, respectively) (38). DBP has low glass transition and melting transition temperatures (-95 °C and -35 °C, respectively) (38, 39). BEHS also has a low melting transition of -67 °C (38).

To choose appropriate solvents, swelling experiments were performed on the cross-linked polybutadiene elastomer. A stoichiometric mixture of MA5 and R45 was completely cured without any solvent. This cross-linked polybutadiene was then swollen at 25 °C in solvents with varied Hansen solubility parameters. The dependence of the mass uptake of solvent on the solvent solubility parameter is illustrated in Figure 2. The maximum network swelling occurs with solvents that have a solubility parameter near 18 MPa$^{1/2}$. BEHS, cyclohexane, and toluene sit at the maximum in the swelling curve. However, cyclohexane and toluene are too volatile for a broad temperature performing gel, while BEHS has a high boiling point. DBP also has a high boiling point but BEHS exhibits higher swelling than DBP, indicating that BEHS is a better solvent for this polybutadiene network. The relative uncertainty in the swelling is ± 5 mass percent, which was the typical standard deviation based on 5 swelling samples per solvent.

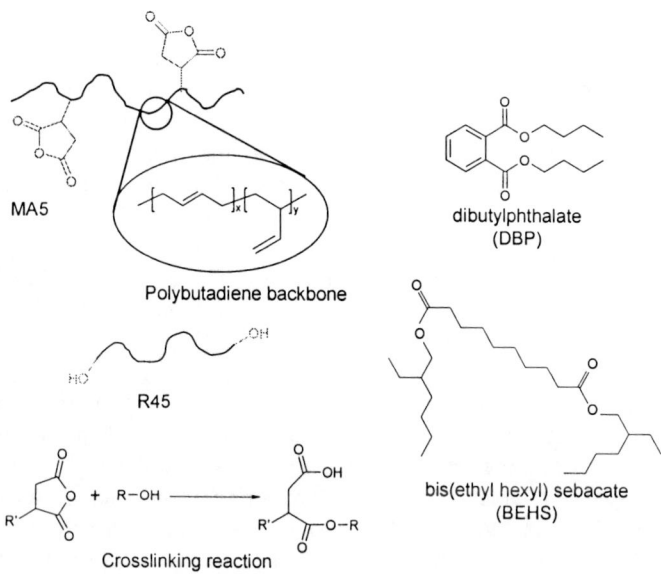

Figure 1. The primary constituents of the gel formulation are shown, as well as the cross-linking reaction.

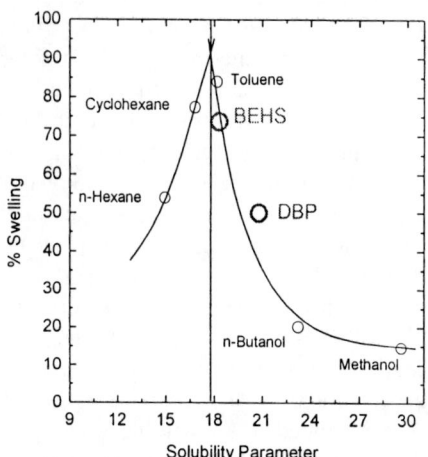

Figure 2. Swelling of cross-linked polybutadiene in solvents with varying solubility parameters. Solubility parameter has units of $MPa^{1/2}$.

Another important issue is to optimize the cross-link density. A higher cross-link density material will offer more stable properties and more robust processing due to the larger number of functional groups per polymer chain. However, a higher cross-link density material will also have a higher modulus and less swelling capability. A lower cross-link density system will provide a more flexible material with better low temperature performance, but can be susceptible to aging or contamination issues due to the low number of functional groups per polymer chain. To assess the impact of cross-link density in these gels, stoichiometric formulations were made with both the MA10 monomer and the MA5 monomer mixed with R45 and 60 mass percent BEHS. MA10 has an average of 5 maleic anhydride groups per polymer chain, while MA5 has an average of 2.5 groups per chain, resulting in an MA5 gel that has a lower cross-link density when compared to the MA10 based gel. Figure 3a plots the shear storage modulus, G', as a function of temperature for these two gel materials. The glass transition temperature (T_g) is similar for both gels at approximately -85 °C, indicating that the cross-link density is low, and the T_g is dominated by the polybutadiene polymer between the cross-linking junctions. As expected, the plateau value for the modulus of the MA5 gel is nearly an order of magnitude lower than the value for the MA10 gel, due to the lower cross-link density of the MA5 based gel. At -70 °C, the MA5 gel has a storage modulus near 10^3 Pa, which is less than half that of the MA10 gel. Figure 3b shows the loss tangent (loss modulus/storage modulus) for the two gel formulations. The loss tangent represents the ability of a system to dissipate energy. The loss tangent spectra are similar for the MA5 and MA10 gel formulations, except in the region between -10 and -70 °C. In that region, the loss tangent is higher for the MA5-based gel. This enhanced energy dissipation combined with a lower modulus value at these low temperatures will enhance the low temperature adhesive performance of the MA5 gel compared with the MA10 gel. Due to this potential for improved low temperature performance, and the fact that the MA5 gel exhibited a reliable and reproducible cure, MA5 was chosen as the best monomer for future gel materials. The relative uncertainty in the modulus values is less than ± 10%, as was demonstrated by measurements on multiple samples of gel at a particular solvent loading.

While BEHS is a better solvent than DBP for the cross-linked polybutadiene (Figure 2), DBP is still a viable candidate solvent due to its low volatility. More information is needed to assess the impact of solvent quality on the gel mechanical properties. Figure 4a plots the storage modulus as a function of temperature for the cross-linked MA5 polybutadiene gel with various mass percent loadings of BEHS. As the BEHS loading increases from 0 to 60% (with increments of 10%), the plateau value for the modulus in the rubbery region decreases by approximately an order of magnitude, and the gel T_g decreases by approximately 25 °C. With the higher solvent loadings of 50 and 60%, the modulus of the BEHS gels exhibits a plateau value at temperatures above -40 °C. However, the gel material is still flexible with a modulus near 10^5 Pa, even at -70 °C. The relative uncertainty in the modulus values is less than ± 10%, as was demonstrated by measurements on multiple samples of gel at a particular solvent loading.

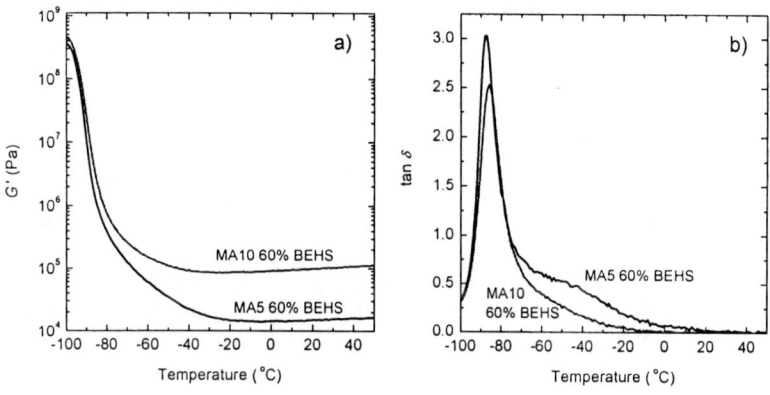

Figure 3. a) Storage modulus as a function of temperature for MA5 and MA10 gel formulations with 60 mass percent BEHS, b) loss tangent as a function of temperature for the same gel formulations.

Figure 4b plots G' as a function of temperature for a stoichiometric MA5 gel formulation with various mass loadings of DBP (also in increments of 10% from 0 to 60%). Similar to BEHS gels, the plateau value of the modulus decreases by an order of magnitude as the solvent loading increases from 0 to 60%. Quantitatively, the decrease in the plateau value of the modulus with increasing solvent loading is the same for both BEHS- and DBP-based gels. However, the T_g decrease when changing solvent loading from 0 to 60% for the DBP-based gels is only 5 °C, compared to 25 °C decrease in T_g with BEHS-based gels. With high DBP loadings 50 and 60%, the plateau value for the modulus occurs at temperatures above -10 °C, while the gels maintain flexibility (a modulus of 10^5 Pa or less) down to approximately -40 °C. Mechanically, both DBP and BEHS gels have a much broader temperature performance range than hydrogels, as the material maintains flexibility at temperatures well below 0 °C. No solvent loss was observed in the DBP based gels, even after a month of aging at 90 °C and several months aging at 60 °C. Solvent loss studies with the BEHS-based gels are ongoing; however, with the similar low volatility of the BEHS solvent, we expect similar results. The BEHS gels with high solvent loadings (Figure 4a) maintain flexibility to temperatures about 30 °C less than DBP-based gels (Figure 4b) with the same solvent loadings. A potential explanation for the difference in lower temperature performance between the DBP- and BEHS-based gels is differences in polymer-solvent miscibility. At elevated temperatures, and the temperature of the curing reaction (75 °C), both DBP and BEHS are miscible with the polymer at high solvent loadings, leading to similar decreases in the plateau value of the modulus with increasing solvent loadings. If DBP is less miscible than BEHS at lower temperatures, this could explain the smaller shift in T_g with the increasing DBP loading compared to the BEHS behavior.

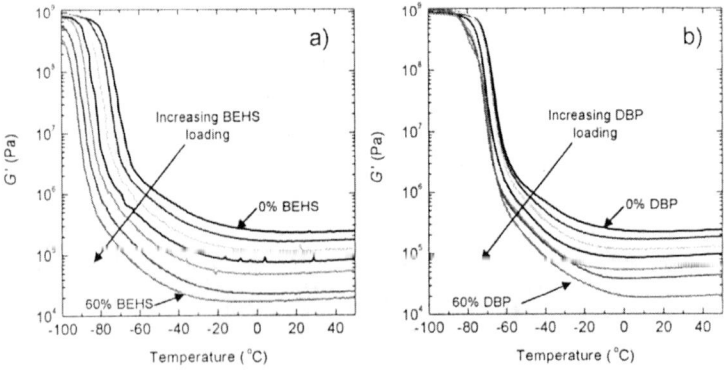

Figure 4. a) Storage modulus as a function of temperature for MA5 gels with various mass loadings of BEHS, b) Storage modulus as a function of temperature for MA5 gels with various mass loadings of DBP.

Qualitatively, the room temperature swelling in Figure 2 illustrates that BEHS is more miscible in the cross-linked polybutadiene network than DBP. The rheology data in Figure 4 also indicates phase separation in the DBP-based gels at lower temperatures. To quantify the upper critical solution temperature (UCST) for these gels, neutron scattering experiments were performed on stoichiometric gel formulations with either deuterium-labeled BEHS at 60 mass percent loading or deuterium labeled DBP at 60 mass percent loading. Figure 5a plots the SANS scattering intensity, I(q), as a function of q, ($q = 2\pi\sin\theta/\lambda$ where θ is the scattering angle), the momentum vector normal to the incident neutron beam, for gels with deuterated DBP. At 15 and 23 °C, the scattering intensity is low thorough the entire q range. At 10 and 5 °C, a dramatic upturn in the low q scattering is observed, indicating microphase separation of the solvent at the lower temperatures. The UCST for the 60%

DBP-based gels is between 10 and 15 °C. Upon phase separation, concentration fluctuations in the gel will diverge to larger length scales, resulting in higher scattering intensity at low q (q is inversely proportional to a length scale). Figure 5b shows USANS data from a gel with 60% BEHS at temperatures from 0 to -100 °C. The coherent scattering is very weak and is temperature independent within the limits of experimental uncertainty, supporting the assertion that microphase separation does not occur in the BEHS gel system. (Data below $q = 5 \cdot 10^{-4}$ Å^{-1} are not plotted for clarity because subtraction of the open beam scattering resulted in large uncertainties.)

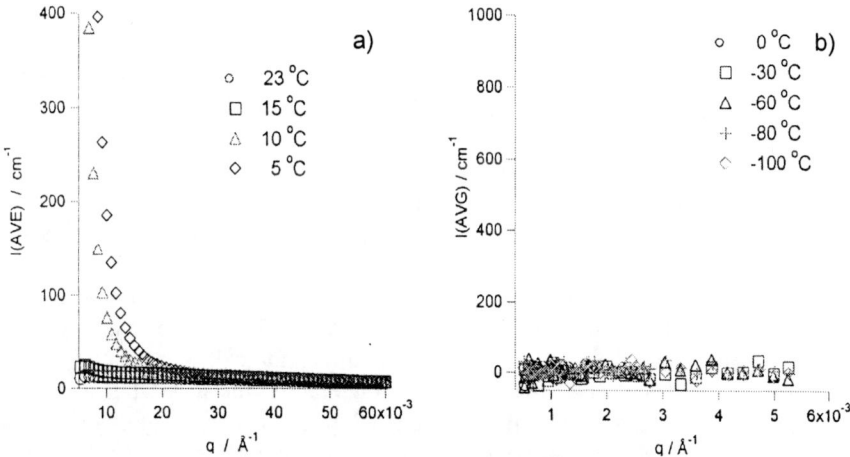

Figure 5. a) SANS from an MA5 gel with 60 mass % deuterium labeled DBP, b) USANS from an MA5 gel with 60 mass % deuterium labeled BEHS.

The solvent quality and resulting polymer-solvent phase behavior had a profound impact on the adhesive performance of the gel materials at low temperatures. Figure 6a shows the tack adhesion at 24 °C for polybutadiene gels with 60 mass percent loading of either DBP or BEHS. For both solvents, the lower cross-link density MA5 gel exhibited higher tack adhesion. Also, at room temperature (where both solvents are miscible with the polymer), the tack adhesion was independent of the solvent type. Figure 6b shows the tack adhesion at -60 °C, where BEHS is miscible in the polymer, but DBP is not. The tack adhesion was much higher with the BEHS gels compared to the DBP-based gels at low temperatures. When the DBP-based gels phase separate at low temperatures, a skin of solvent is excluded towards the gel surface, which degrades the tack adhesion.

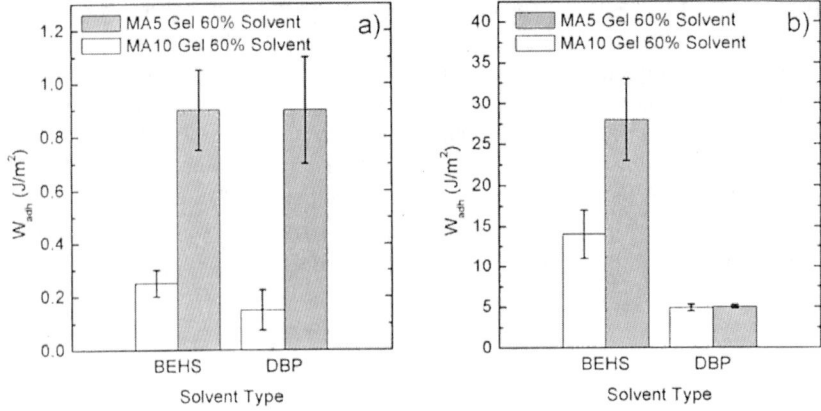

Figure 6. a) tack adhesion for gels at 24 °C, b) tack adhesion for gels at -60 °C.

The tack adhesion in Figure 6 qualitatively correlates with the rheological loss tangent spectra in Figure 3b. At -60 °C, the loss tangent is higher than at 24 °C for both the MA10 and MA5 gels, and the tack adhesion is also significantly higher for both these gels at 60 °C. In addition, the loss tangent at -60 °C for the MA5 gel is higher than the MA10 gel, and the resulting tack adhesion also higher for the MA5 gel at this temperature. While the correlation between tack adhesion and loss tangent is qualitative, the loss tangent can be used to assess whether a gel will have high tack performance. Ongoing research is focused on a more quantitative understanding linking the gel rheology and tack adhesion, which can be extended across a variety of materials, and utilized in a materials design methodology.

Conclusions

A non-aqueous, polymer gel was developed that maintains stable, mechanical, adhesive, and equilibrium performance over a temperature range from -70 to 70 °C. The gel was composed of two polybutadiene monomers, one contained maleic anhydride functional groups, the other contained hydroxyl groups. The maleic anhydride/hydroxyl cross-linking reaction was performed in the presence of high loadings of a solvent, BEHS, resulting in a low cross-link density polybutadiene network that was highly swollen. Swelling experiments, rheological measurements, neutron scattering, and tack adhesion were utilized to characterize the gel material and illustrated that the solvent quality had an important impact on the temperature performance range of the gel.

Acknowledgment

Sandia is a multiprogram laboratory operated by Sandia Corporation, a Lockheed Martin Company, for the United States Department of Energy under contract DE-AC04-94AL85000. We acknowledge the support of the National Institute of Standards and Technology, U.S. Department of Commerce, in providing the neutron research facilities used in this work. This work utilized facilities supported, in part, by the National Science Foundation under Agreement No. DMR-9986442.

References

1. Dusek, K. (ed.) *Responsive Gels: Volume Phase Transitions, Advances in Polymer Science,* **1993**, 109, (Springer, Berlin). H. Ito, *Jpn. J. Appl. Phys.* **1992**, *31*, 4273.
2. Flory, P. J. *Principles of Polymer Chemistry,* **1953**, 576, Cornell University Press, Ithaca, NY.
3. de Gennes, P. G. *Scaling Concepts in Polymer Physics,* **1979**, 137, Cornell University Press, Ithaca, NY.
4. Tanaka, T. *Phys. Rev. Lett.* **1978**, *40*, 820.
5. Tanaka, T. *Phys. Rev. A* **1978**, *17*, 763.
6. Hamlen, R. P.; Kent, C. E.; Shafer, S. N. *Nature* **1965**, *206*, 1149.
7. Steinberg, I. Z.; Oplatka, A.; Katchalsky, A. *Nature* **1966**, *210*, 568.
8. Sussman, M. V.; Katchalsky, A. *Science* **1970**, *167*, 45.
9. Kwon, I. C.; Bae, Y. H.; Kim, S. W. *Nature* **1991**, *354*, 291.
10. Miyata, T.; Asami, N.; Uragami, T. *Nature* **1999**, *399*, 766.
11. Murdan, S. *J. Controlled Release* **2003**, *92*, 1.
12. Peppas, V.; Huang, V.; Torres-Lugo, M.; Ward, J. H.; Zhang, J. *Annu. Rev. Biomed. Eng.* **2000**, *2*, 9.
13. Gallegos, C.; Franco, J. M. *Curr. Opin. Coll. Interface Sci.* **1999**, *4*, 288.
14. Lenhart, J. L., Sun, W.-Q.; Payne, G. F. *Chem. Eng. Sci.* **1997**, *52*, 645.
15. Wilson, M. J.; Fuchs, A.; Gordaninejad, F. *J. Appl. Polym. Sci.* **2002**, *84*, 2733.
16. Jackson, D. K.; Leeb, S. B.; Narvaez, P.; Fusco, D.; Lubton, E. C. Jr., *IEEE Trans. Indust. Electr.* **1997**, *44*, 217.
17. Beebe, D. J.; Moore, J. S.; Baure, J. M.; Yu, Q.; Liu, R. H.; Devadoss, C.; Jo, B.-H. *Nature* **2000**, *404*, 588.
18. Liu, R. H.; Yu, Q.; Beebe, D. J. *J. Microelectromech. Sys.* **2002**, *11*, 45.
19. Irvin, D. J.; Goods, S. H.; Whinnery, L. L. *Chem. Mater.* **2001**, *13*, 1143.
20. Kubota, N.; Watanabe, H.; Konaka, G.; Eguchi, Y. *J. Appl. Polym. Sci.* **2000**, *76*, 12.

21. Abraham, K. M.; Alamgir, M. J. *Power Sources* **1993**, *43*, 195.
22. Otake, M.; Kagami, Y.; Inaba, M.; Inoue, H. *Robotics Autom. Sys.* **2002**, *40*, 185.
23. Wang, G.; Kuroda, K.; Enoki, T.; Grosberg, A.; Masamune, S.; Oya, T.; Takeoka, Y.; Tanaka, T. *Proc. Nat. Acad. Sci.* **2000**, *97*, 9861.
24. Jiang, H.; Su, W.; Mather, P. T.; Bunning, T. J. *Polymer* **1999**, *40*, 4593.
25. Chang, S-C.; Yang, Y. *Appl. Phys. Lett.* **1999**, *75*, 2713.
26. Weissman, J. M.; Sunkara, H. B.; Tse, A. S.; Asher, S. A. *Science* **1996**, *274*, 959.
27. Chang, S.-C.; Yang, Y. *Appl. Phys. Lett.* **1999**, *75*, 2713.
28. Hu, Z.; Chen, Y.; Wang, C.; Zheng, Y.; Li, Y. *Nature* **1998**, *393*, 149.
29. Holtz, J. H.; Asher, S. A. *Nature* **1997**, *389*, 829.
30. Li, J.; Hong, X.; Liu, Y.; Li, D.; Wang, Y.; Li, J.; Bai, Y.; Li, T. *Adv. Mater.* **2005**, *17*, 163.
31. Al-Sharji, H. H.; Grattoni, C. A.; Dawe, R. A.; Zimmerman, R. W. *Oil Gas Sci. Technol.* **2001**, *56*, 145.
32. Hirokawa, Y.; Tanaka, T. *J. Chem. Phys.* **1984**, *81*, 6379.
33. Park, T. G. *Biomaterials* **1999**, *20*, 517.
34. Schmaljohann, D.; Oswald, J.; Jorgensen, B; Nitschke, M.; Beyerlein, D.; Werner, C. *Biomacromolecules* **2003**, *4*, 1733.
35. Elaissari, A.; Rodrigue, M,; Meunier, F.; Herve, C. *J. Magn. Magn. Mater.* **2001**, *225*, 127.
36. MA5 and R45 were obtained from Sartomer. See Sartomer product sheets at www.sartomer.com.
37. Billmeyer, F. W. Jr., *Textbook of Polymer Science* **1984**, John Wiley & Sons, New York, NY.
38. Materials Safety Data Sheets for Dibutylphthalate and bis(ethylhexyl)sebacate.
39. Wang, L.-M.; Velikov, V.; Agnell, C. A. *J. Chem. Phys.* **2002**, *117*, 10184.

Chapter 7

Detection of Toxic Chemicals for Homeland Security Using Polyaniline Nanofibers

Shabnam Virji[1,2], Richard B. Kaner[2], and Bruce H. Weiller[1]

[1]Materials Processing and Evaluation Department, Space Materials Laboratory, The Aerospace Corporation, Mail Stop M2-248, P.O. Box 92957, Los Angeles, CA 90009-2957
[2]Department of Chemistry and Biochemistry and California NanoSystems Institute, University of California at Los Angeles, Los Angeles, CA 90095-1569

The electrical properties of the conducting polymer polyaniline change greatly upon exposure to various chemicals. Specifically, polyaniline undergoes doping and dedoping chemistry with acids and bases that result in conductivity changes of over eight orders of magnitude. This large range in conductivity can be utilized to make polyaniline chemical sensors. Polyaniline nanofibers are chemically synthesized using a simple, template-free method that produces nanofibers with narrow size distributions. They are easily cast on microelectrode arrays and are shown to respond significantly better than conventional films to a number of different gases, such as acids, bases, hydrazine, and organic vapors. This is explained through their high surface area, small diameter, and porous nature of the nanofiber films that appear to allow better diffusion of vapors into the films. Polyaniline nanofibers disperse well in water and, as a result, have been used to make new composite materials with water soluble compounds, such as metal salts. These composite

© 2008 American Chemical Society

films can then be used to enhance sensing to gases that unmodified polyaniline would otherwise not be able to detect. For example, metal salt/polyaniline nanofiber composite films are used to enhance polyaniline's ability to sense hydrogen sulfide due to a reaction of the metal salt with hydrogen sulfide that releases a strong acid that then dopes the polyaniline, resulting in a significant increase in conductivity. The wide range in gas detection capabilities and the use of composite films makes polyaniline nanofibers versatile chemical sensor materials that have good potential for many chemical detection applications, including homeland security.

Introduction

There is a need for sensors to detect toxic chemicals for homeland security purposes. The sensors need to be small, low-powered, and have a fast time response. Many of the toxic chemicals that must be detected are strong acids or bases. Polyaniline has great potential to detect these chemicals because of its strong interaction with acids and bases that lead to large conductivity changes.

Polyaniline, a conducting polymer, has been widely studied for a number of different applications because of its simple, and reversible, acid doping/base dedoping chemistry (*1*). Polyaniline can undergo a transition from its insulating emeraldine base form to its conducting emeraldine salt form that can lead to over ten orders of magnitude change in conductivity. These changes in conductivity can be used to make chemical sensors. Previous work on polyaniline sensors includes detection of acids (*2*), bases (*3*), organic solvents (*4*), and redox active agents (*5*).

Conventional polyaniline has been previously used as a chemical sensor but has been limited in its sensitivity and time response. Previous work on enhancing the detection capabilities of polyaniline has included the use of thinner films (*6*). The disadvantages of this method are the loss of robustness of the film and the difficulty in making such thin films with good control. Another way to increase sensitivity to gases is to change the morphology of the polyaniline. In particular, nanostructured forms of polyaniline, such as nanowires, nanofibers, or nanorods, have recently received much attention because their small diameters are expected to allow for fast diffusion of gas molecules into the structures (*7*).

Polyaniline nanostructures have been synthesized with specific structural-directing materials added to the polymerization bath including templates (*8, 9*) or functional molecules (*10, 11*). Both of these methods lead to complex synthetic conditions that require the removal of the templates and/or produce nanostructures with small yields and low reproducibility. Electrospinning is a method that does not require a template, but only certain materials can be made on a limited scale by this method due to the difficulty controlling different experimental parameters such as the applied voltage and solution viscosity (12). Recently, we have developed a new, simple, synthetic method to make polyaniline nanofibers with nearly uniform diameters between 30 and 120 nm, with lengths varying from 500 nm to several microns. This method is template-free and is based on a modification of the classic chemical oxidative polymerization of aniline. In this method, the aniline is polymerized at the interface between the two-phases of an organic-aqueous system instead of using the traditional homogeneous aqueous solution of aniline, acid, and oxidant (13).

Polyaniline nanofibers have recently received much attention as sensors because they have many advantages over conventional films. In particular, we have designed polyaniline nanofiber sensors for toxic gases and have shown that they give much faster and larger responses due to their small fiber diameters and high surface areas (2, 5, 13). Subsequently, others have used arrays of oriented polyaniline nanofibers (14) as well as single nanowires (7) to detect gases.

In this paper, we present an overview of our work on polyaniline nanofiber chemical sensors. Specifically, we show how polyaniline nanofibers outperform conventional polyaniline as chemical sensors upon exposure to hydrochloric acid (HCl), ammonia (NH_3), hydrazine (N_2H_4), and hydrogen sulfide (H_2S). We will also give examples of how the sensor response can be enhanced by using polyaniline nanofiber composite films.

Experimental

Polyaniline nanofibers were chemically synthesized in an aqueous-organic, two-phase system and purified by dialysis (13). A typical synthesis consists of first dissolving aniline in an organic solvent and dissolving ammonium peroxydisulfate and camphorsulfonic acid (CSA) in water. The two solutions are carefully added together to form an interface between the two layers and, after 3-5 min, the entire water phase is filled with green, polyaniline nanofibers. The products are purified by dialysis against deionized water. The emeraldine base form of the nanofibers is made by dialysis using 0.1 M ammonium hydroxide. The final product exists as a dispersion in water and is used to cast films for the sensors. The transmission and scanning electron microscope images of the nanofibers are shown in Figure 1.

The emeraldine salt form of conventional polyaniline was chemically synthesized from aniline by oxidative polymerization using ammonium peroxydisulfate in an acidic media (1). Washing the salt form with 0.1 M ammonium hydroxide produces the emeraldine base form of polyaniline. Conventional polyaniline solutions were prepared in hexafluoroisopropanol (HFIP), at a concentration of 2 mg/mL, or in *N*-methylpyrrolidinone (NMP), at a concentration of 1 mg/mL.

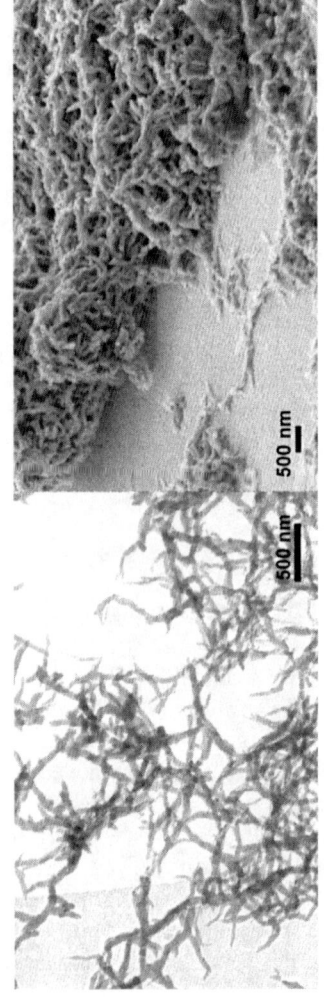

Figure 1. Transition electron microscope (left) and scanning electron microscope (right) images of polyaniline nanofibers.

The sensor films were made by drop casting a fixed amount of polyaniline solution onto planar electrodes using a disposable microliter pipette. The planar interdigitated electrode geometry consists of 50 pairs of gold fingers on glass or oxidized silicon, each finger having dimensions of 10 μm × 3200 μm × 0.18 μm (width × length × height) and a 10 μm gap between fingers. The array sensor consists of 6 different interdigitated electrode sensors fabricated on one die. The substrates are mounted in standard ceramic packages and wire bonded via gold contact pads.

Film thicknesses were measured with a profilometer (DekTak II). Electron microscope images were obtained using a field-emission scanning electron microscope (JEOL JSM-6401F). Electrical resistances (DC) were measured with a programmable electrometer (Keithley 617). A low-current scanner card and switch system (Keithley 7158/7001) was used to multiplex measurements over ten sensors from two sensor arrays. Instruments were controlled and read by computer using a GPIB interface and LabView software.

Gas exposures used certified gas mixtures of 100 parts-per-million (ppm) HCl in nitrogen (Scott Specialty Gases), 100 ppm NH_3 in nitrogen (Matheson), or 200 ppm H_2S in nitrogen (Scott Specialty Gases). For hydrazine exposures, a certified permeation tube (KinTek) of hydrazine was used with a calibrated emission rate. The concentration was determined using a well-known colorimetric method with *p*-dimethylaminobenzaldehyde (*15*). Mass flow controllers were used to control separate flows of nitrogen gas and the calibrated gas mixture. A bubbler was used to generate humidity and was measured directly using a humidity sensor (Vaisala).

Results and Discussion

Polyaniline can undergo a reversible acid/base doping/dedoping process. In the emeraldine salt form, polyaniline is conductive. This form can then be converted to the insulating emeraldine base form of polyaniline, upon exposure to base as in Equation 1, where HX is a protonic acid:

Emeraldine Base   Emeraldine Salt

Various forms of doped polyaniline have been widely used to produce efficient ammonia sensors (*3*). However, the dedoped emeraldine base form has not been thoroughly studied as an acid sensor (*16*). One of the main problems associated with producing an acid sensor is the measurement of the baseline conductivity of the insulating form of polyaniline. This conductivity is very low, on the order of 10^{-10} S/cm. Most electronic devices are not capable of measuring such low conductivities. The instrumentation used to measure all of the acid sensor data in this paper has a very large dynamic range and allows us to make such measurements.

Measuring acid concentration is important for many applications, specifically for space and homeland security applications. Hydrochloric acid is released from the exhaust plume of solid rocket motors that use ammonium perchlorate as the rocket propellant. This plume can be dangerous in high concentrations and monitoring its concentration and dispersion are important tasks. Hydrochloric acid and other acid gases like nitric acid and hydrofluoric acid are also important and very dangerous toxic industrial chemicals and need to be monitored because of their low permissible exposure limits.

Figure 2a shows the response of conventional and nanofiber polyaniline thin films to 100 ppm HCl. The left axis is the normalized resistance, that is, the time dependent resistance divided by the initial resistance (R/R_0), and the bottom axis is the time in seconds. Upon exposure to the acid, polyaniline undergoes a conversion from its insulating emeraldine base form to its conducting emeraldine salt form and we observe orders of magnitude decrease in resistance within a few seconds. Both the conventional and nanofiber films undergo large resistance decreases; however, the nanofiber film outperforms the conventional film by orders of magnitude. Even with a nanofiber film that is almost 3 times thicker, it still outperforms the conventional film by many orders of magnitude.

We have shown previously that the response of the nanofiber film is thickness independent, whereas, the conventional film is thickness dependent (2). That is, no matter how thick the nanofiber film is, it responds similarly to the analyte gas. However, for conventional polyaniline, thicker films give a slower and smaller response. This is an important property of the nanofiber sensors because we do not need to control the thickness of the film. A likely explanation of this property is that the nanofiber film is very porous while the conventional film is a dense film. The gas can, therefore, penetrate more easily through the nanofiber film than the conventional film. Also, the gas needs only diffuse into the 50 nm fiber as opposed to penetrating a micron thick film that will take much more time.

In addition to acids, polyaniline also responds well to bases. Figure 2b shows the response of the emeraldine salt form of polyaniline to ammonia. The resistance increases because polyaniline undergoes a transition from the emeraldine salt (conducting form) to the emeraldine base (insulating form). This conversion leads to a resistance increase of about two orders of magnitude for the nanofiber film. This response is much smaller and slower than the response to acid because of a different mechanism associated with each reaction. Upon exposure to base, a deprotonation step occurs removing the acid. This step takes longer than protonation so the response is slower. The deprotonation step is likely slower because, upon exposure of the emeraldine salt form to

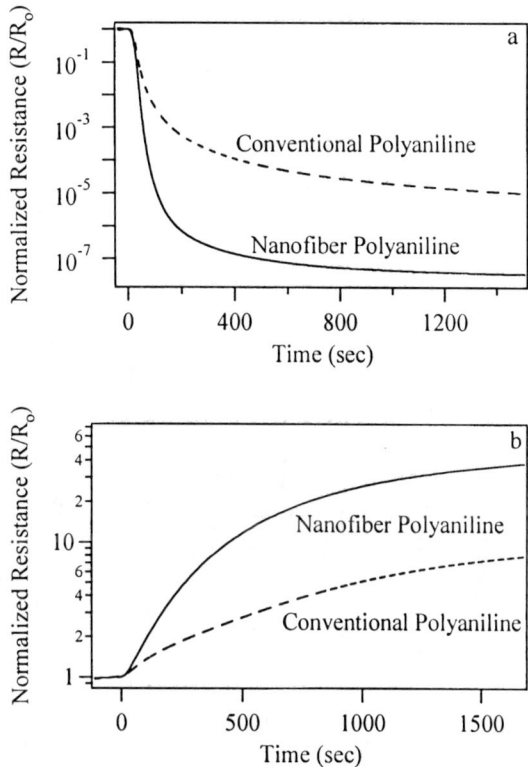

Figure 2. Comparison of nanofiber (— 2 μm) and conventional (--- 0.7 μm) polyaniline films exposed to a) 100 ppm HCl, b) 100 ppm NH_3.

ammonia, NH_4Cl is formed which is in equilibrium with HCl and NH_3. The HCl is therefore not fully removed when using low concentrations of gaseous NH_3, and complete dedoping is not achieved. The difference between the nanofiber and conventional film is again related to the higher porosity, surface area, and small diameters of the nanofibers. Figure 3 shows that nanofibrillar polyaniline has good reversibility upon exposure to ammonia.

The response time is another important parameter of a sensor. The response time (τ_{90}) is defined as the time it takes to reach 90% of the full value. Figure 4 shows the response time of polyaniline upon exposure to acid. The graph shows that a nanofiber film responds much faster than a conventional film, even though the nanofiber film is twice as thick as the conventional film. This is also related to the porosity of the film and small diameter of the nanofibers.

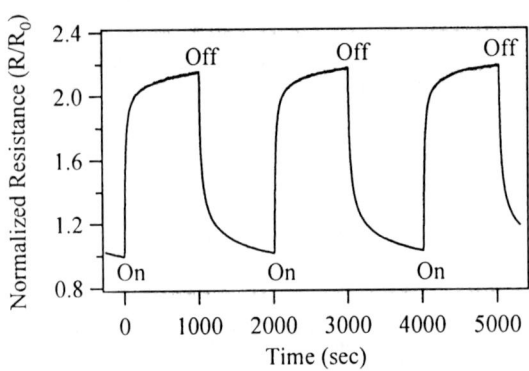

Figure 3. Reversibility of polyaniline nanofibers films exposed to NH_3 (6 ppm) at 45% relative humidity.

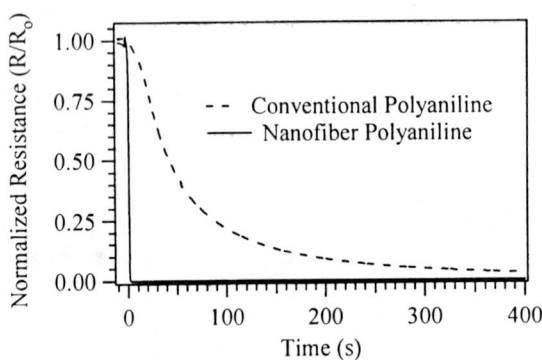

Figure 4. Response time of nanofiber (— 2.0 μm, τ_{90} = 3 sec) and conventional (--- 1.0 μm, τ_{90} = 200 sec) polyaniline films exposed to HCl (100 ppm).

Hydrazine detection is important because it is a carcinogen with a threshold limit value (TLV) of only 10 ppb. Hydrazine, monomethylhydrazine (MMH), and unsymmetrical dimethylhydrazine (UDMH) are rocket fuels, explosive, and highly toxic. Previous work on hydrazine sensing has been carried out utilizing other conducting polymers including polypyrrole and polythiophene (5). Polythiophene sensors are very sensitive because they can detect hydrazine on the parts-per-billion (ppb) level, but a problem is their instability in air. Polypyrrole sensors are more stable, but the hydrazine detection limit is only ~1%, orders of magnitude higher than the threshold limit value.

Polyaniline can exist in other oxidation states including the leucoemeraldine form (fully reduced) and the pernigraniline form (fully oxidized) (17). It is well known that hydrazine can reduce polyaniline (5). Polyaniline nanofibers are good for sensing hydrazine because, unlike the other conducting polymers, they are air stable and undergo an oxidation state change upon exposure to hydrazine. Figure 5 shows the response of nanofiber polyaniline and conventional polyaniline to 3 ppm hydrazine. Note that the left nanofiber and right conventional axes are different. The graph shows a resistance increase indicative of conversion from a conducting state to an insulating state. However, unlike exposure to ammonia, the film color becomes white, consistent with conversion to the leucoemeraldine form. This is further verified with reflectance-IR measurements on the films (5). It is also noteworthy that the conventional film shows a very small resistance increase, on the order of 25%, while the nanofiber film shows a signal change of over 30 times that of the conventional film.

A method to make more sensitive sensors is to create composite materials that can increase the response of polyaniline to the analyte. One type of material that we have used to enhance the response of *conventional* polyaniline to hydrazine are fluoroalcohols, specifically hexafluoroisopropanol (HFIP) (5). Polyaniline nanofibers dissolve in HFIP and lose their nanofibrillar morphology, so nanofibers cannot be examined with this solvent. Figure 6 shows the response of a polyaniline film with added HFIP and a film with added N-methylpyrrolinone (NMP). The HFIP film shows a dramatic resistance decrease of over four orders of magnitude while the NMP film shows a small resistance increase of about 25%. This is the same NMP film as that shown in Figure 5.

The data in Figure 6 can be explained by the reaction of the fluoroalcohol with hydrazine to produce HF and byproducts as given in Equation 2. The HF dopes the polyaniline to convert the insulating emeraldine base form to the conducting emeraldine salt form as given in Equation 1. In addition to HFIP, its derivative, hexafluoro-2-phenylisopropanol, shows a similar response. There is no prior reference to this reaction in the literature, although hydrazine has been used to defluorinate carbon nanotubes (18). Furthermore, a strong exothermic reaction is observed upon the addition of HFIP to aqueous hydrazine resulting in a pH drop from 11 to 3 (19):

$$N_2H_4 + (CF_3)_2CHOH \rightarrow HF + \text{byproducts} \qquad (2)$$

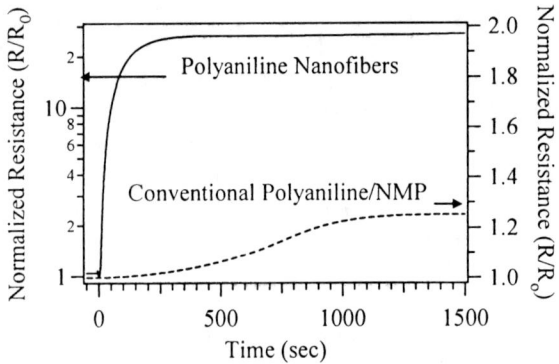

Figure 5. Comparison between nanofiber (— 2 μm) and conventional (--- 0.7 μm) polyaniline films exposed to 3 ppm N_2H_4. The left axis is for nanofiber polyaniline and the right axis is for conventional polyaniline.

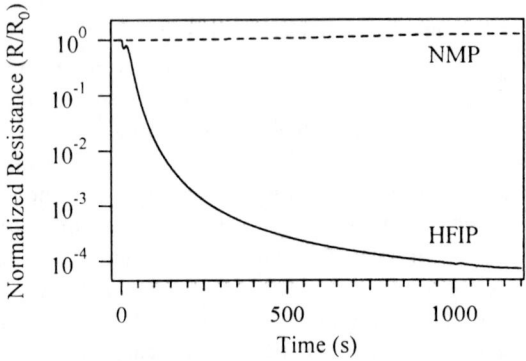

Figure 6. Comparison between NMP (--- 0.3 μm) and HFIP (— 0.3 μm) conventional polyaniline films exposed to 3 ppm N_2H_4.

The concept of using composites with polyaniline can be used to detect other gases, such as hydrogen sulfide (H_2S). Hydrogen sulfide is a very weak acid with a pK_a of 7 and it cannot dope polyaniline directly. About 10^4 times as much weak acid is required to dope polyaniline when compared to a strong acid (20). One way to enhance the sensitivity of polyaniline to H_2S is to use inorganic–organic composite materials.

We have recently developed a method to detect H_2S using a composite film of polyaniline with transition metal salts (21). The metal salt reacts with the H_2S to form a metal sulfide and a strong acid (Equation 3). The strong acid then dopes the polyaniline resulting in a conversion to the conducting emeraldine salt form (Equation 1) where MS is a metal sulfide:

$$MCl_2 + H_2S \rightarrow MS + 2HCl \qquad (3)$$

Figure 7 shows the response of polyaniline modified with copper chloride ($CuCl_2$) and unmodified polyaniline to 10 ppm H_2S. The unmodified film shows essentially no change in resistance, while the modified polyaniline film shows a resistance decrease of over four orders of magnitude.

We have also shown that other transition metal chlorides, in addition to copper chloride, can be used and each metal chloride gives a different response depending on the solubility product constant (K_{sp}) of the resulting metal sulfide. Figure 8 shows the response of polyaniline composite films to the different metal chlorides. As can be seen from the graph, the response is dependent on the metal chloride used. This can be correlated to the solubility product constant of the resulting metal sulfide (Table I).

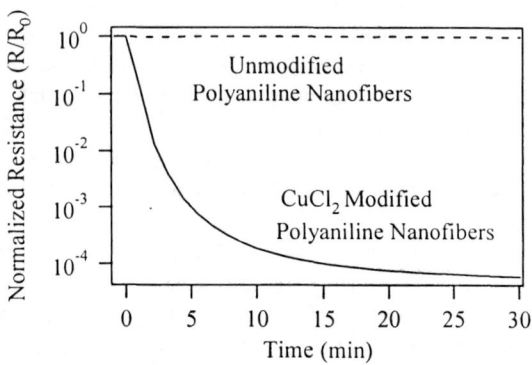

Figure 7. Comparison between unmodified nanofiber (--- 0.4 μm) and $CuCl_2$ modified (— 0.4 μm) polyaniline films exposed to 10 ppm H_2S. (Reproduced with permission from reference 21. Copyright 2005 Wiley-VCH.)

Table I. Solubility product constants for metal sulfides

Metal Sulfide	Solubility Product Constant (K_{sp})[a]
ZnS	2×10^{-4}
CdS	8×10^{-7}
CuS	6×10^{-16}

[a] CRC Handbook of Chemistry and Physics, 85th edition.

The solubility product constant (K_{sp}) is a dissociation constant for metal complexes in aqueous solution. A smaller K_{sp} value means that the complex is less likely to dissociate in water and implies that the complex is more stable. More stable metal sulfides drive Equation 3 to form the acid which then dopes the polyaniline resulting in a large resistance decrease in the film. Table I shows the K_{sp} values for different metal sulfides. According to the table, CuS is the most stable metal sulfide and ZnS is the least stable. This is consistent with CuS exhibiting the largest response, whereas ZnS gives the smallest response.

It is noteworthy that this reaction only occurs in the presence of polyaniline. Films of the neat chloride show no formation of sulfides. We believe that this is due to coordination of the metal cation to the amine nitrogens of the polyaniline backbone, which makes the cations more available to react with H_2S. This is consistent with the observation that copper acetate can react directly with H_2S (22).

In conclusion, nanofiber films outperform conventional polyaniline films with significantly better performance in both sensitivity and time response due to their high surface area, small nanofiber diameter, and porous nature of the films. In addition to being able to detect a number of different chemicals directly, polyaniline composite materials can also be used to enhance detection for gases that would not be possible to detect otherwise. We have shown that for conventional polyaniline, composites with fluoroalcohols enhance hydrazine detection due to a reaction between e.g., hydrazine and hexafluoroisopropanol. The acid product, HF, in turn, dopes polyaniline resulting in an over four order of magnitude decrease in resistance. We have also demonstrated that new composite materials formed from metal salts and polyaniline nanofibers show enhanced response to hydrogen sulfide. The proposed mechanism for this response is the reaction of H_2S with metal salts to form the corresponding metal sulfide and an acid, which then dopes the polyaniline resulting in over four orders of magnitude decrease in resistance. The response is correlated with the K_{sp} of the resulting metal sulfide. The more stable metal sulfides, with smaller K_{sp} values, show a larger response to hydrogen sulfide.

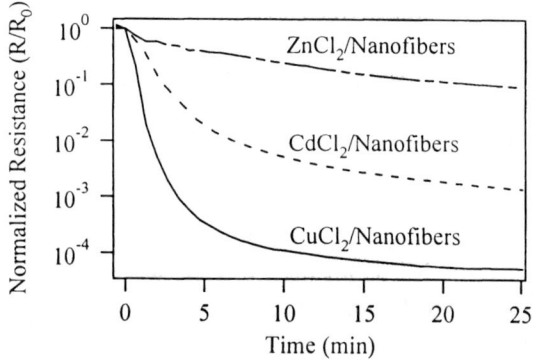

Figure 8. Resistance changes of polyaniline nanofiber films containing $ZnCl_2$ (— - -), $CdCl_2$ (---), and $CuCl_2$ (—) on exposure to hydrogen sulfide (10 ppm). The relative humidity was 45% and all films were 0.25 μm thick. (Reproduced with permission from reference 21. Copyright 2005 Wiley-VCH.)

To further examine these materials, analyte concentrations will be varied to determine the ranges of response and the limits of detection. Other work in progress is geared toward the measurement of nanofiber responses with varying humidity and temperature environments. In addition, diffusion studies are planned in order to more fully understand the interaction of analytes with the nanofibers. We are also working on developing new composite materials with polyaniline nanofibers in order to control the response of conducting polymers to chemicals of interest for these and other new applications.

Acknowledgments

This work has been supported by the Homeland Security Advanced Projects Research Agency (HSARPA) contract #HSHQPA-05-9-0036, The Aerospace Corporation's Independent Research and Development Program (B.H.W.), the Microelectronics Advanced Research Corp. (R.B.K.) and an NSF-IGERT fellowship (S.V.). We would like to thank Mr. Jesse Fowler for the design and fabrication of the sensor array and Dr. Jiaxing Huang, Dr. Dan Li and Ms. Christina Baker for helpful discussions and contributions to this work.

References

1. Huang, W.-S.; Humphrey, B. D.; MacDiarmid, A. G. *J. Chem. Soc, Faraday Trans.* **1986**, *82*, 2385-2400.

2. a) Huang, J.; Virji, S.; Weiller, B. H.; Kaner, R. B. *Chem. Eur. J.* **2004**, *10*, 1314-1319. b) Virji, S.; Huang, J.; Kaner, R. B.; Weiller, B. H. *Nano Lett.* **2004**, *4*, 491-496. c) Huang, J.; Virji, S.; Weiller, B. H.; Kaner, R. B. *J. Am. Chem. Soc.* **2003**, *125*, 314-315.
3. a) Domansky, K.; Baldwin, D. L.; Grate, J. W.; Hall, T. B.; Li, J.; Josowicz, M.; Janata, *Anal. Chem.* **1998**, *70*, 473-481. b) Kukla, A. L.; Shirshov, Y. M.; Piletsky, S. A. *Sens. Act. B* **1996**, *37*, 135-140. c) Athawale, A. A.; Chabukswar, V. V. *J. Appl. Poly. Sci.* **2001**, *79*, 1994-1998.
4. a) Miller, L. L.; Bankers, J. S.; Schmidt, A. J.; Boyd, D. C. *J. Phys. Org. Chem.* **2003**, *13*, 808-815. b) Tan, C. K.; Blackwood, D. J. *Sens. Act. B* **2000**, *71*, 184-191.
5. a) Virji, S.; Kaner, R. B.; Weiller, B. H. *Chem. Mater.* **2005**, *17*, 1256-1260. b) Ellis, D. L.; Zakin, M. R.; Bernstein, L. S.; Rubner, M. F. *Anal. Chem.* **1996**, *68*, 817–822. c) Ratcliffe, N. M. *Anal. Chim. Acta* **1990**, *239*, 257-262.
6. a) Agbor, N. E.; Cresswell, J. P.; Petty, M. C.; Monkman, A. P. *Sens. Act. B* **1997**, *41*, 137-141. b) Granholm, P.; Paloheimo, J.; Stubb, H. *Syn. Met.* **1997**, *84*, 783-784.
7. Liu, H.; Jun, K.; Czaplewski, D. A.; Craighead, H. G. *Nano Lett.* **2004**, *4*, 671-675.
8. Wu, C. G.; Bein, T. *Science* **1994**, *264*, 1757-1759.
9. Martin, C. R. *Acc. Chem. Res.* **1995**, *28*, 61-68.
10. Yu, L.; Lee, J. I.; Shin, K. W.; Park, C. E.; Holze, R. *J. Appl. Poly. Sci.* **2003**, *88*, 1550-1555.
11. Liu, J. M.; Yang, S. C. *Chem. Comm.* **1991**, 1529-1531.
12. MacDiarmid, A. G.; Jones, W. E.; Norris, I. D.; Gao, J.; Johnson, A. T.; Pinto, N. J.; Hone, J.; Han, B.; Ko, F. K.; Okuzaki, H.; Llanguno, M. *Syn. Met.* **2001**, *119*, 27-30.
13. a) Huang, J.; Kaner, R. B. *J. Am. Chem. Soc.* **2004**, *126*, 851-855. b) Huang, J.; Kaner, R. B. *Angew. Chemie International Ed.* **2004**, *43*, 5817-5821.
14. Liu, J.; Lin, Y.; Liang, L.; Voigt, J. A.; Huber, D. L.; Tian, Z. R.; Coker, E.; Mckenzie, B.; Mcdermott, M. J. *Chem. Eur. J.* **2003**, *9*, 604-611.
15. Schmidt, E. W. *Hydrazine and Its Derivatives*, John Wiley & Sons: New York, NY, 1984.
16. Koul, S.; Dhawan, S. K.; Chandra, S. *Indian J. Chem.* **1997**, *36a*, 901-904.
17. Zeng, X.-R.; Ko, T.-M. *Polymer* **1998**, *39*, 1187-1195.
18. Mickelson, E. T.; Huffman, C. B.; Rinzler, A. G.; Smalley, R. E.; Hauge, R. H.; Margrave, J. L. *Chem. Phys. Let.* **1998**, *296*, 188-194.
19. Virji, S.; Kaner, R. B.; Weiller, B. H. *Polym. Prepr. (Am. Chem. Soc. Div. Polym. Chem.)* **2005**, *1*, 624-627.

20. Hatchett, D. W.; Josowicz, M.; Janata, J. *J. Phys. Chem. B* **1999**, *103*, 10992-10998.
21. Virji, S.; Fowler, J. D.; Baker, C. O.; Huang, J.; Kaner, R. B.; Weiller, B. H. *Small* **2005**, *1*, 624-627.
22. Virji, S.; Kaner, R. B.; Weiller, B. H., *Inorg. Chem.* **2006**, *45*, 10467-10471.

Chapter 8

Applications of Nanoparticles in Scintillation Detectors

Suree S. Brown, Adam J. Rondinone, and Sheng Dai[*]

Chemical Sciences Division, Oak Ridge National Laboratory, P.O. Box 2008, Oak Ridge, TN 37831-6201

Applications of commercially available, highly efficient inorganic scintillators (particle size: μm) are limited by their low solubilities in both polymeric and sol-gel matrices. On the other hand, organic scintillators, though highly soluble in polystyrene-based matrices, are not compatible with an efficient neutron inorganic absorber, ^{6}Li, and their applications with ^{6}Li as neutron scintillators are strictly limited. Here, preparation and surface modification of organic nanoparticles and inorganic nanocrystals are demonstrated as a means to increase dispersion and compatibility of scintillators with neutron-absorbing materials and matrices. A survey of nanoparticles, including PPO, POPOP-doped polystyrene nanoparticles, CdSe/ZnS core/shell quantum dots, Y_2O_3:Ce (5%), $LaPO_4$:Ce (10%), and $^6Li_3PO_4$ nanocrystals, in various matrices, along with their results in beta, alpha, or neutron detection is discussed.

Introduction

Advanced radiation detectors, especially for neutron and gamma radiation, are important for detecting "dirty" bombs, monitoring U-235 and Pu-239 for nuclear arms control/verification, nuclear smuggling, special nuclear material (SNM) detection, and waste characterization, as well as for fundamental research in nuclear physics, solid-state physics, chemistry, biology, and neutron radiography. One of the oldest, and still the most useful, methods for detecting ionizing radiation (e.g., alpha particles, beta particles, gamma rays) is through the scintillation process. Scintillation occurs when the energy of ionizing radiation is absorbed by certain crystalline inorganic or organic materials, resulting in the emission of UV-vis light from the absorbing materials (*1*). The key to the development of advanced radiation detectors lies in the synthesis and characterization of efficient scintillation materials.

In the case of neutron detection, the conversion of incident neutrons into detectable charged particles is normally required. Of particular importance in the detection of slow neutrons, *ca.* below 0.5 eV, are $^{10}B(n,\alpha)$ and $^{6}Li(n,\alpha)$ reactions. We report that by taking advantage of the high Q-value (i.e., large excess energy imparted to charged particles) of the ^{6}Li nuclear reaction ($^{6}Li + {}^{1}n \rightarrow {}^{3}H + {}^{4}He$, Q-value = 4.78 MeV), we can detect slow neutrons. The energy of alpha particle produced as a secondary radiation is deposited directly onto scintillation materials.

The ideal scintillation material should meet the following criteria: high scintillation efficiency, wide range of linear energy conversion, transparency to the wavelength of its own emission, short emission decay time, good optical quality, index of refraction near that of glass (~1.5) (*2*), emission spectrum matched to the spectral sensitivity of the detector, and high durability (*3*). Commercially available inorganic scintillation materials possess many of the desired properties; however, their particle sizes in the range of microns lead to the opacity of scintillation detectors and, consequently, lowered scintillation efficiencies.

During the past two decades, there have been extensive investigations on inorganic nanocrystals, especially semiconductor nanocrystals or quantum dots (QDs). With appropriate surface functional groups, nanocrystals can be dispersed in various solvents, sol-gels, or polymers, yielding transparent products. In the case of QDs, quantum confinement effect (*4–7*) offers the plausibility of a better matching between the emission spectrum and the sensitivity of a photomultiplier tube (PMT) or other light sensors used. Here, a survey of representative organic nanoparticle and inorganic nanocrystals embedded in various matrices, along with their results in beta, alpha, or neutron detection, is presented. The ultimate goal of this study is to be able to identify the most suitable nanoparticle-based scintillation materials for specific radiation detection purposes.

A better understanding of how certain nanoparticle-based scintillators behave under radiation interactions is also expected to be gained.

Experimental Section

Reagents and Syntheses

All reagents used were of highest grades commercially available. Most were used as received, except styrene and methylstyrene, which were distilled freshly prior to usage. All syntheses and handling were carried out under an inert atmosphere. 2,5-diphenyloxazole (PPO), 1,4-bis-2-(5-phenyloxazolyl)-benzene (POPOP)-doped polystyrene (PS) and polyvinyltoluene (PVT) nanoparticles were synthesized by ultrasonication of micelle solutions of styrene and methylstyrene, respectively, under conditions modified from the procedures reported by Biggs and Grieser (8). CdSe/ZnS core/shell quantum dots were prepared by a selected procedure published previously (9). Cerium-doped yttrium oxide (Y_2O_3:Ce) nanocrystal samples were prepared by three different methods. The detailed synthesis by the first method, Pechini-type in-situ polymerizable complex (IPC) method, has been reported (10). This method utilizes polyesterification between metal complexes of an α-hydroxycarboxylic acid (e.g., DL-malic acid) and a polyhydroxy alcohol (e.g., ethylene glycol), followed by calcination, yielding homogeneous Y_2O_3:Ce. The second method employed the slow decomposition of urea and hexamethylenetetramine (HMTA) at elevated temperatures (up to 235 °C), providing an in-situ generated controlled amount of ammonia. The third method involved the formation of complexes between the polyethylene glycol, PEG-8000, as a chelating agent and the yttrium and cerium metals. Heat treatment of yttrium, cerium-PEG complexes led to the formation a mixed-metal oxide. Cerium-doped lanthanum phosphate ($LaPO_4$:Ce) nanocrystals were prepared by a modified method from the detailed synthesis reported by Haase and co-workers (11–13). Surface modification of nanocrystals was done by injecting a large excess of mixed surfactants to the reaction mixture at 100 °C. A mixture of dodecylamine (DDA) and bis(2-ethylhexyl) hydrogenphosphate (BEHP) and a mixture of oleylamine (OA) and BEHP were compared in this study. The surface modification was allowed to proceed for 1 h at 100 °C before the reaction mixture was worked up by precipitation with methanol. Synthesis of lithium-6 phosphate ($^6Li_3PO_4$) from Li-6 oleate (95% ^6Li) for neutron detection was modified from the synthesis of $LaPO_4$:Ce nanocrystals.

Commercially available, highly efficient liquid and solid scintillators were tested as standards for comparison with our samples. Ultima Gold™ (PerkinElmer) is a standard liquid scintillation counting cocktail, containing PPO (ca. 1 wt %) in a mixture of aromatic compounds, and phosphate and succinate surfactants. BC-400™ (Bicron) is a plastic scintillator composed of organic fluors, PPO and POPOP, at <3 wt % in PVT. KG2™ (Bicron) is a cerium-activated lithium silicate glass scintillator, containing 7.5 wt % of Li (95% ^{6}Li) as a neutron absorber, and Ce^{3+} as a scintillation material.

Radiation Detection

Instrumentation was assembled for comparing pulse-height spectra from each experimental detector. The pulse-height analysis system consisted of a one-inch-diameter photomultiplier tube (Hamamatsu R1924A, with E2924-500 socket assembly, custom-built housing) followed by amplifiers (Ortec 113 preamplifier and 575A shaping main amplifier) and a computer-interfaced multichannel analyzer (Spectrum Techniques UCS-20). The electronics were interconnected with 93 Ω or 52 Ω coaxial cable, as appropriate. A 4.7 kΩ resistor was installed across the preamplifier input because the PMT socket assembly lacked an anode load resistor as supplied. A double-layer film-handling bag surrounded the PMT assembly to maintain light-tight conditions, while facilitating rapid exchange of the detector being tested.

Pulse height spectra for alpha bombardment were collected by affixing an alpha source to the end of the detector retention tube, facing the detector, at approximately 1 cm distance. The source used was a 1.0 μCi ^{241}Am disk source mounted in the end of a ThorLabs optics tube. Counts were accumulated for a live time of 250 s. The MCA conversion gain was set at 256 channels.

The procedure for neutron measurements involved placing the detector near a neutron source storage drum. Measurements were made with a 3 Ci AmBe neutron source and a 3 mCi ^{252}Cf spontaneous fission neutron source housed in 55-gallon drums. The drum contained a thick neutron moderator/absorber shield, and, thus, it was assumed that thermal neutrons predominated in the energy distribution outside the drum. The detector and PMT were shielded from fission gamma radiation by 5-cm thick lead bricks, and an additional 3-cm thick, high-density polyethylene moderator was interposed between the bricks and the source drum. Pulse-height spectra were accumulated over 1.5×10^4 s for each sample.

After collection, the pulse-height spectra were analyzed with the MCA software supplied by Spectrum Techniques. Peak centroid channel, full-width-at-half-maximum (FWHM) and net and gross peak areas were obtained. The gain setting of the shaping main amplifier was also recorded for each test.

Results and Discussion

In order to match the light emission of scintillation materials with the spectral sensitivity of PMT, nanoparticles and inorganic nanocrystals with the emission in the range of 350–450 nm were synthesized, characterized, and applied to radiation detection. Results from the detection of beta particles, alpha particles, and neutron radiation are presented here.

Beta Particle Detection by PPO, POPOP-Doped Polystyrene Nanoparticle Liquid Scintillator

PPO and POPOP are among the most commonly used organic primary and secondary fluors, respectively. Despite their high efficiencies in both liquid and plastic scintillators, one limitation is their incompatibility with hydrophilic reagents. In liquid scintillation techniques, an efficient extraction of radioactive nuclides into the organic phase where PPO and POPOP dissolve is often mandatory. In neutron detection by plastic scintillators, the use of an efficient, yet hydrophilic, neutron absorber, ^{6}Li, is strictly limited if not prohibited.

Figure 1. PS nanoparticles (left) and PPO, POPOP-doped PS nanoparticles (right) (a) under room light (b) under UV light.
(See page 1 of color inserts.)

Transparent, colloidal, aqueous dispersions of PS and PVT nanoparticles (10 wt %, average diameter: 63 nm, polydispersity: 0.3) embedded with PPO (1 wt %) and POPOP were synthesized and examined as an approach to overcome this chemical incompatibility. Figure 1a shows transparent aqueous dispersions containing PS nanoparticles under room light. Under a UV light source (Figure 1b), bright photoluminescence was observed in the PPO, POPOP-doped PS nanoparticle sample (on the right) while very dim photoluminescence was observed from an undoped PS nanoparticle sample (on the left). Considering the very low solubility of PPO and POPOP in aqueous solutions, this confirms the presence of organic scintillators inside nanoparticles.

The transparent PPO, POPOP-doped PS nanoparticle-aqueous dispersion was tested for beta particle detection by adding 5.0 mL of ^{14}C ethanolic solution to 0.5 mL of the scintillation solution. The result from nanoparticle dispersion was compared to that of a standard liquid scintillation cocktail, Ultima Gold^{TM}, and a background count in the absence of ^{14}C (Table I). The disintegration count from PPO, POPOP-doped PS nanoparticle dispersion was well above the background count, indicating the detection of beta particles from ^{14}C–without false alarm. The low count compared to that of Ultima Gold^{TM} indicated quenching in this aqueous sample. This count was comparable to that of an organic scintillator medium with a 34% quenching, measured previously (14).

Table I. Detection of Beta Particles from ^{14}C by Liquid Scintillators

Sample	dpm
Ultima Gold^{TM}	1010072
PPO, POPOP in PS nanoparticles	6000.69
Background count	8.17

Even in the absence of covalent bonding between organic scintillators and PS or PVT matrices, these PPO, POPOP-doped nanoparticle-colloidal dispersions were found to have long-term stability, which may be attributed to small particle sizes. The incorporation of PPO, POPOP-PS or PVT nanoparticles into lithiated sol-gels as solid-state neutron detectors is also an on-going study by our group. In general, the principle demonstrated here may also be applicable to other organic scintillators.

Alpha Particle Detection by Nanocrystalline Inorganic Scintillators

Many inorganic nanocrystals, including CdSe/ZnS core/shell quantum dots and nanocrystals doped with Ce^{3+} as luminescent centers (e.g., Y_2O_3:Ce, $LaPO_4$:Ce), were synthesized as described above. With particles sizes below 5 nm, narrow size distribution, and high crystallinity, they demonstrated high light yields, a desirable property for efficient scintillators.

CdSe/ZnS core/shell quantum dots (QDs)

A host of literature is available on the preparation and properties of CdSe and highly luminescent CdSe/ZnS core/shell QDs (*15-19*). The passivation of CdSe surface defect sites by a robust inorganic shell increased the photoluminescence quantum yields to as high as 50%. Here, highly luminescent CdSe/ZnS core/shell QDs with the maximum emission at 472 nm were embedded at 50 wt % in a polystyrene matrix. A thick, transparent film of QDs in polystyrene on a quartz disc was then examined for the detection of alpha particles from a [241]Am source. Figure 2 is the pulse height spectrum showing a peak at a very low relative pulse height (ca. 0.10 channel #/gain).

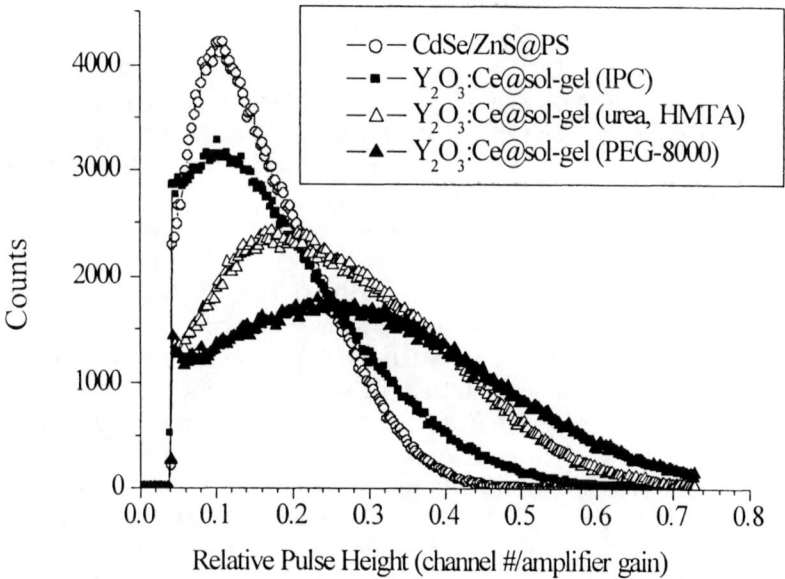

Figure 2. Pulse height spectra from alpha particle detection by CdSe/ZnS core/shell and $Y_{1.90}Ce_{0.10}O_3$ nanocrystals prepared by different methods.

Ce-doped Y_2O_3 nanocrystals (Y_2O_3:Ce)

Y_2O_3 nanocrystals at the dopant (Ce^{3+}) concentration of 5% were found to be highly luminescent (*14*). All $Y_{1.90}Ce_{0.10}O_3$ nanocrystals prepared by three different methods emitted at 401 nm. Thick films of 50 wt % of nanocrystals in sol-gels on quartz discs were tested for the detection of alpha particles from a ^{241}Am source and their pulse height spectra are shown in Figure 2. The relative pulse heights of all $Y_{1.90}Ce_{0.10}O_3$ nanocrystal samples did not vary significantly (i.e., from 0.10 to 0.27 channel #/gain), which is to be expected from the same identity of scintillation materials. Their relative pulse heights were also very low and not significantly different from that of CdSe/ZnS core/shell QDs. On the other hand, the peak from the QD sample was obviously more resolved than those from $Y_{1.90}Ce_{0.10}O_3$ nanocrystal samples.

Ce-doped lanthanum phosphate nanocrystals (LaPO$_4$:Ce)

Different from most doped inorganic nanocrystals, lanthanide phosphates can be prepared with almost a stoichiometric amount of the dopant added during the synthesis (*11*). Recent studies also confirmed a very efficient doping of a specific lanthanide ion, Eu^{3+}, as occupying both interior and surface sites of LaPO$_4$ nanocrystals (*20, 21*). In this study, the LaPO$_4$:Ce nanocrystals with the efficient doping of Ce^{3+} were also expected due to the nearly identical ionic radii of La^{3+} and Ce^{3+}, and a 10 mol % dopant was found to be an optimal concentration for the highest emission efficiency (*14*).

$La_{0.90}Ce_{0.10}PO_4$ nanocrystals were surface modified with mixed surfactants: a mixture of DDA and BEHP or a mixture of OA and BEHP. The surface modified nanocrystals were dispersed at 50 wt % in polystyrene, forming translucent thick films on quartz discs. Pulse height spectra from alpha particle detection (Source: ^{241}Am) of nanocrystal samples superimposed with the result from a highly efficient plastic scintillator, BC-400^{TM} are presented in Figure 3.

The relative pulse heights of the peaks from $La_{0.90}Ce_{0.10}PO_4$ nanocrystal samples at 0.25 and 0.45 channel #/gain were still much lower than that of BC-400^{TM} (at 2.55 channel #/gain). However, their peaks showed much higher counts and were better resolved (i.e., much lower FWHM) than the peak from BC-400^{TM}. The lower pulse height from the nanocrystals surface capped with OA and BEHP, compared to that of DDA and BEHP-capped nanocrystals, is likely related to color quenching caused by the yellowish OA present in the sample. Pulse height spectra (not presented) of 10, 25, and 50 wt % samples of $La_{0.90}Ce_{0.10}PO_4$-DDA, BEHP-capped nanocrystals in polystyrene also showed increased counts with the increasing wt % of nanocrystals at the same relative pulse height of the peaks (*14*). This behavior of $La_{0.90}Ce_{0.10}PO_4$ scintillation nanocrystals is in accord with the linearity observed in bulk inorganic scintillators.

Compared to the peaks from CdSe/ZnS core/shell QDs and $Y_{1.90}Ce_{0.10}O_3$ nanocrystal samples, the peaks from $La_{0.90}Ce_{0.10}PO_4$ nanocrystal samples showed the highest relative pulse heights, highest counts, and similar resolution to the peak from the QD sample (FWHM: ca. 0.25 channel #/gain). Due to different

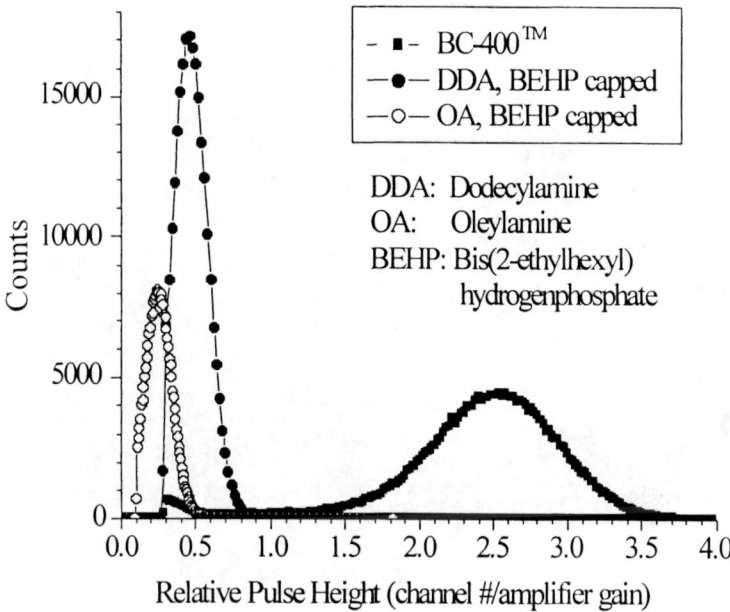

Figure 3. Pulse height spectra from alpha particle detection by $La_{0.90}Ce_{0.10}PO_4$ nanocrystals with different surface groups compared with BC-400^{TM}.

optical quality of nanocrystalline samples, it is still inconclusive as to what the most important determining factors are in order to obtain high relative pulse heights of the detection peaks. Among results presented here, $La_{0.90}Ce_{0.10}PO_4$ nanocrystals seemed to show the most desired scintillation properties in alpha particle detection.

Slow Neutron Detection with Li-6 Phosphate ($^6Li_3PO_4$) Nanocrystals as Neutron Absorbers

Using a similar chemical reaction to the synthesis of lanthanide phosphate nanocrystals, $^6Li_3PO_4$ can be prepared in high yields at much faster growth rates. Figure 4 is a TEM image of $^6Li_3PO_4$ nanocrystals, showing particle sizes in the range of 5–7 nm. Translucent, thick films of polystyrene containing organic scintillators, PPO (1 wt %) and POPOP, and $^6Li_3PO_4$ nanocrystals (20 and 7 wt %, corresponding to 3 and 1 wt % of ^6Li, respectively) as a neutron-absorbing material were prepared.

Pulse height spectra from the detection of shielded and moderated neutrons (Source: combined 3 Ci AmBe and 3 mCi ^{252}Cf) were superimposed with the

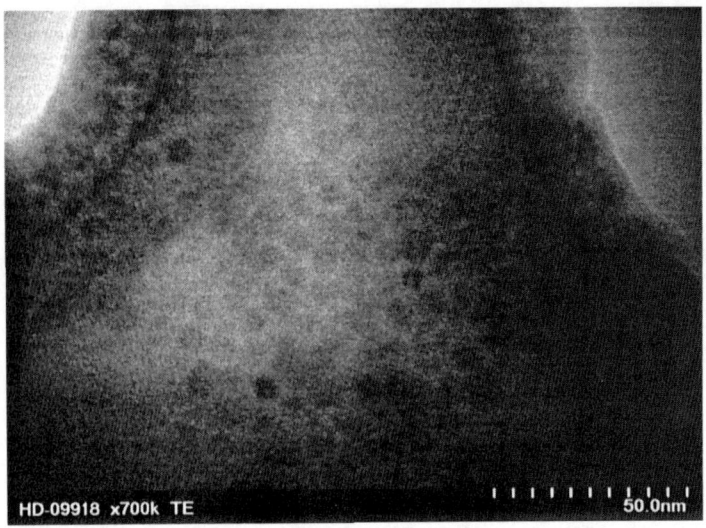

Figure 4. TEM image of $^6Li_3PO_4$ nanocrystals.

result from a highly efficient scintillator, KG2^{TM} in Figure 5. The relative pulse heights of $^6Li_3PO_4$ nanocrystals-containing scintillators were comparable to or higher than that of KG2^{TM}, even though samples were much less transparent. On the other hand, their photon counts were much lower than that of KG2^{TM}. Quenching by a high loading of organic scintillators, PPO and POPOP, might have contributed to the low counts and scintillation samples with lower concentrations of PPO and POPOP are now under an investigation.

Comparing pulse height spectra from $^6Li_3PO_4$ nanocrystals-containing scintillators at 1 and 3 wt % of ^6Li, photon counts increased linearly with the increasing wt % of ^6Li, indicating more neutron events. Accompanied with the increasing photon counts, the relative pulse height (i.e., channel #/gain) also decreased. This, however, is likely to be the result of increasing opacity of the sample as the amount of $^6Li_3PO_4$ nanocrystals increased. Due to a high relative pulse height from the more transparent sample (i.e., 1 wt % ^6Li), the neutron peak was well resolved from another peak at 2.2 channel #/gain. This peak at 2.2 channel #/gain had not been identified, but it could also be a gamma peak.

Further surface modification of $^6Li_3PO_4$ nanocrystals is in progress with the aim to prepare optically clear plastic, neutron scintillators with high % loading of ^6Li. The incorporation of $^6Li_3PO_4$ nanocrystals as neutron absorbers with a nanocrystal scintillator, specifically $La_{0.90}Ce_{0.10}PO_4$, in polystyrene-based matrices as slow neutron detectors is also in progress.

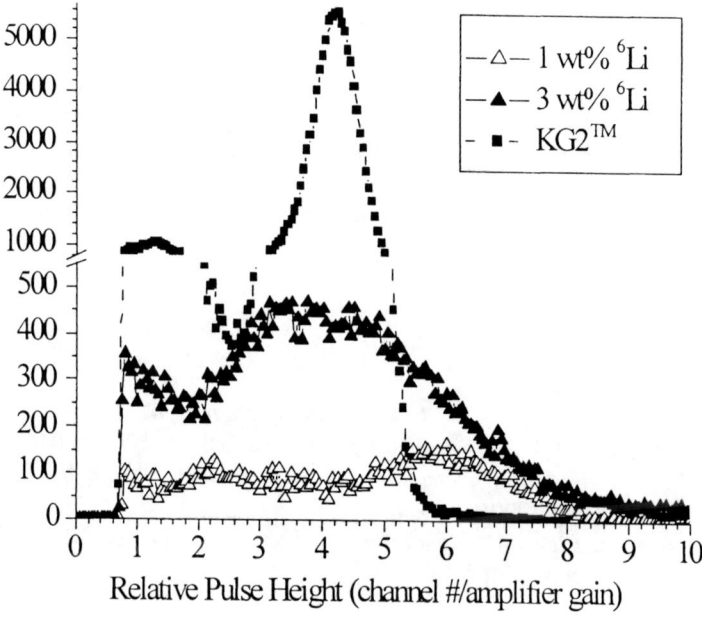

Figure 5. Pulse height spectra from neutron detection by $^6Li_3PO_4$ nanocrystals compared with KG2 ™.

Conclusions

Transparent, aqueous dispersion of PPO, POPOP-doped PS nanoparticles (10 wt %) as a liquid scintillator showed activity well above the background counts in beta particle detection. Thick films of CdSe/ZnS core/shell QDs in polystyrene, $Y_{1.90}Ce_{0.10}O_3$ nanocrystals in sol-gels, and $La_{0.90}Ce_{0.10}PO_4$ in polystyrene showed well defined pulse height spectra in alpha particle detection. Although their relative pulse heights were much lower than that of BC-400™, the most efficient nanocrystal scintillator with the highest pulse height was $La_{0.90}Ce_{0.10}PO_4$. The $La_{0.90}Ce_{0.10}PO_4$ nanocrystal-containing scintillator also showed a much higher resolution and higher photon counts than the corresponding peak from BC-400™. $^6Li_3PO_4$ nanocrystals showed the potential usage as an efficient neutron-absorbing material, with high relative pulse heights in slow neutron detection. The dispersion of representative nanoparticles and nanocrystals in various matrices clearly showed their advantages over micron-sized scintillators. Appropriate surface modification to further increase their dispersion in both sol-gel and polymer matrices is currently pursued as the key to enable this technology.

Acknowledgments

The Oak Ridge National Laboratory is managed for the Department of Energy under contract No. DE-AC05-00OR22725 by UT-Battelle, LLC. Funding for this work is through the support of the Department of Energy NA-22. This research was supported in part by the appointment for S. S. Brown to the ORNL Research Associate Program, administered jointly by ORNL and Oak Ridge Institute for Science and Education.

References

1. L'Annunziata, M. F. In *Handbook of Radiation Analysis*, 2nd ed.; L'Annunziata, M. F., Ed.; Academic Press: San Diego, 2003; p 846.
2. Knoll, G. F. *Radiation Detection and Measurement*, 3rd ed; John Wiley & Sons: New York, 2000; p 219.
3. Miura, N. In *Phosphor Handbook*; Shionoya, S.; Yen, W. M., Eds.; CRC Press: Boca Raton, 1999; pp 523-524.
4. Henglein, A. *Chem. Rev.* **1989**, *89*, 1861.
5. Schmid, G. *Chem. Rev.* **1992**, *92*, 1709.
6. Alivisatos, A. P. *Science* **1996**, *271*, 933.
7. Nirmal, M.; Brus, L. E. *Acc. Chem. Res.* **1999**, *32*, 407.
8. Biggs, S.; Grieser, F. *Macromolecules* **1995**, *28*, 4877.
9. Dai, S.; Saengkerdsub, S.; Im, H.-J.; Stephan, A. C.; Mahurin, S. M. In *Unattended Radiation Sensor Systems for Remote Applications*; Trombka, J. I., Ed.; American Institute of Physics, 2002; pp 220–224.
10. Saengkerdsub, S.; Im, H.-J.; Willis, C.; Dai, S. *J. Mater. Chem.* **2004**, *14*, 1207.
11. Lehmann, O.; Meyssamy, H.; Kömpe, K.; Schnablegger, H.; Haase, M. *J. Phys. Chem. B* **2003**, *107*, 7449.
12. Riwotzki, K.; Meyssamy, H.; Kornowski, A.; Haase, M. *J. Phys. Chem. B* **2000**, *104*, 2824.
13. Riwotzki, K.; Meyssamy, H.; Schnablegger, H.; Kornowski, A.; Haase, M. *Angew. Chem., Int. Ed.* **2001**, *40*, 573.
14. Brown, S. S.; Im, H.-J.; Rondinone, A. J.; Dai, S.; *unpublished results*.
15. Murray, C. B.; Norris, D. J.; Bawendi, M. G. *J. Am. Chem. Soc.* **1993**, *115*, 8706.
16. Peng, X.; Schlamp, M. C.; Kadavanich, A. V.; Alivisatos, A. P. *J. Am. Chem. Soc.* **1997**, *119*, 7019.
17. Peng, Z. A.; Xiaogang, P. *J. Am. Chem. Soc.* **2001**, *123*, 183.

18. Dabbousi, B. O.; Rodriguez-Viejo, J.; Mikulec, F. V.; Heine, J. R.; Mattoussi, H.; Ober, R.; Jensen, K. F.; Bawendi, M. G. *J. Phys. Chem. B* **1997**, *101*, 9463.
19. Hines, M. A.; Guyot-Sionnest, P. *J. Phys. Chem. B* **1996**, *100*, 468.
20. Lehmann, O.; Kömpe, K.; Haase, M. *J. Am. Chem. Soc.* **2004**, *126*, 14935.
21. Haase, M.; Riwotzki, K.; Meyssamy, H.; Kornowski, A. *J. Alloys Compd.* **2000**, *303-304*, 191.

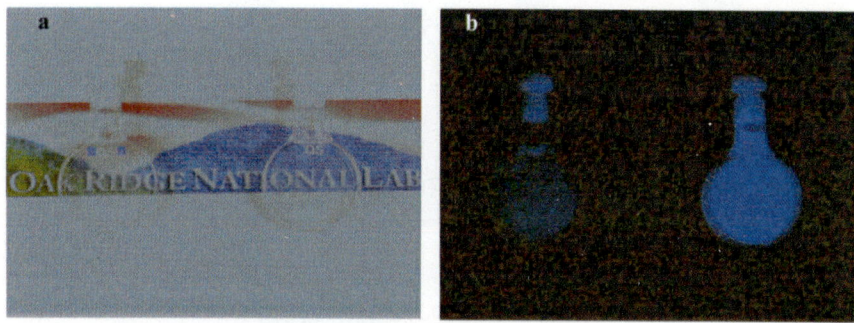

Figure 8.1. PS nanoparticles (left) and PPO, POPOP-doped PS nanoparticles (right) (a) under room light (b) under UV light.

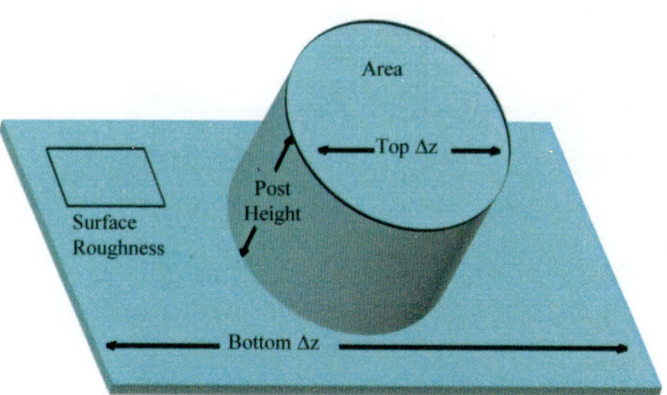

Figure 9.1. Schematic depiction of metrology measurements of the post structures utilized for comparison.

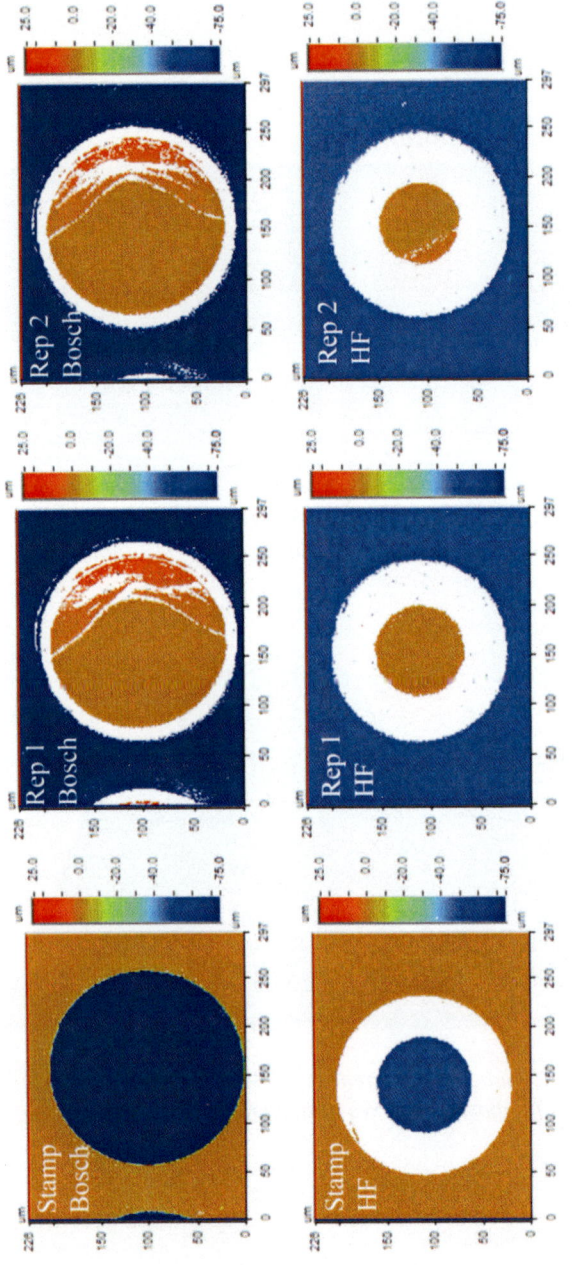

Figure 9.2. Interferometer data taken of stamp and corresponding replicates for 200-μm diameter posts for Bosch and wet HF replication.

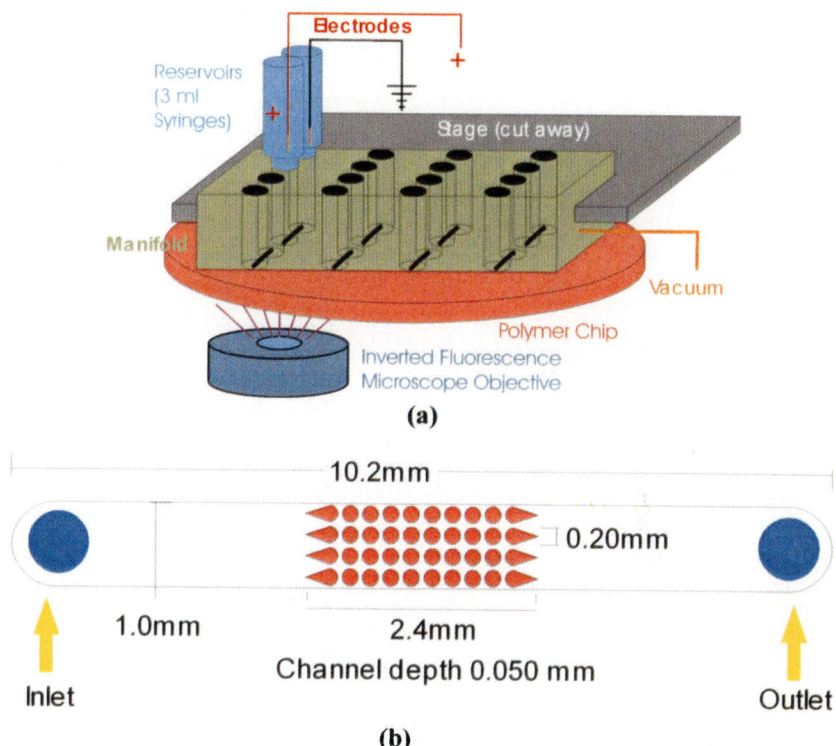

Figure 9.3. Schematic depiction of the experimental setup showing (a) each device contained 8 individual channels on a chip injection molded from Zeonor® 1060R resin. The chip was sealed to a manifold with a vacuum chuck and the apparatus was clamped to a holder on top of the microscope stage. (b) A schematic of a channel with 4 x 10 post configuration.

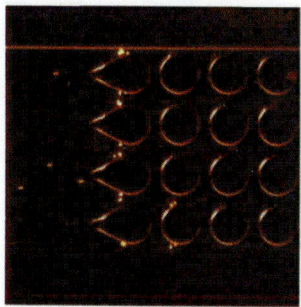

Figure 9.5. Demonstration of separation of polystyrene beads by size. 2-μm red beads trap at 50 V/mm, while 1-μm green beads pass between posts.

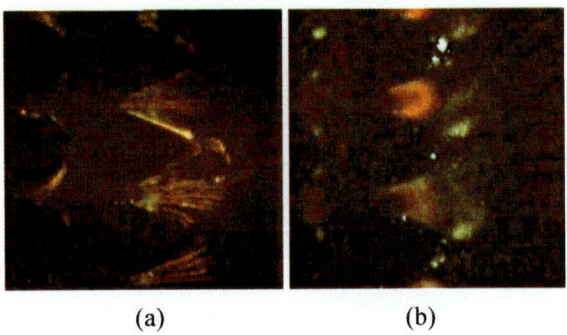

Figure 9.6. Differential trapping comparison of biological tracers. (a) Differential banding of red-labeled vegetative B. subtilis cells and green-labeled B. subtilis spores at 170 V/mm. (b) Differential banding of green-labeled B. subtilis spores and red-labeled B. thuringiensis spores at 200 V/mm.

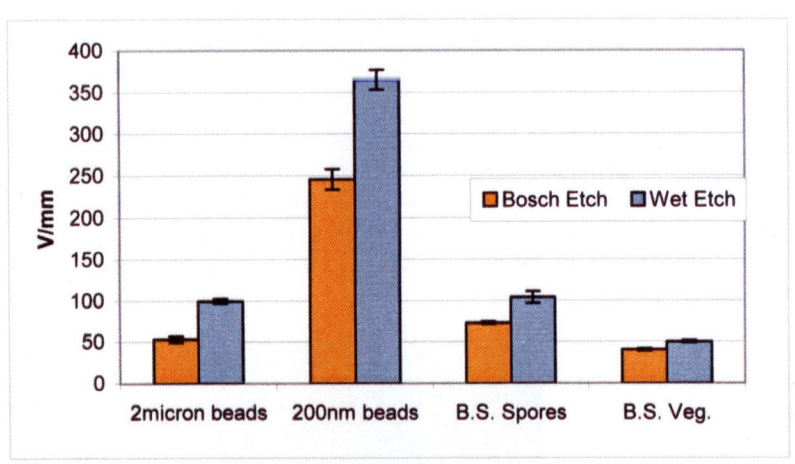

Figure 9.7. Trapping thresholds of particles studied as a function of device fabrication technique. The particle trapping threshold is observed to be a function of particle size, particle conductivity, and morphology.

Figure 9.8. The finite element meshes of the HF (upper) and Bosch (lower) iDEP channels.

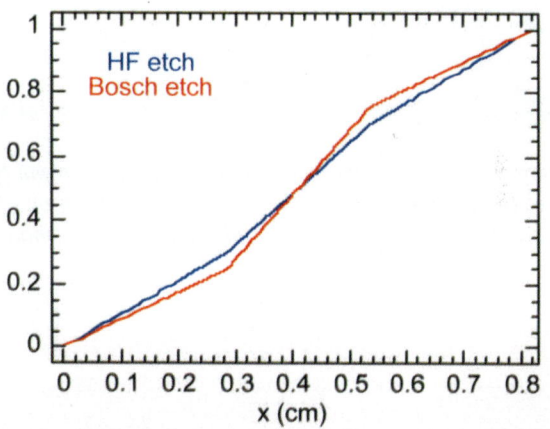

Figure 9.9. The non-dimensional electric potential as a function of distance along the longitudinal centerlines of the Bosch and HF channels.

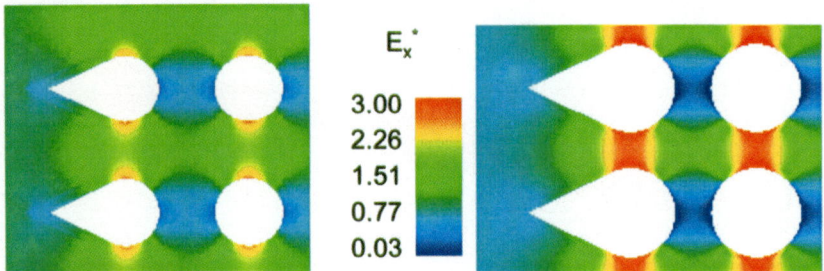

Figure 9.10. The x-component of the non-dimensional electric field in the leading region of the post array for the HF (left) and Bosch (right) channels.

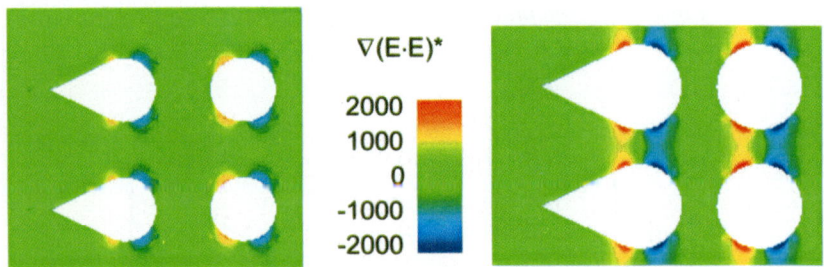

Figure 9.11. The x-component of the non-dimensional gradient of the electric field intensity in the leading region of the post array for the HF (left) and Bosch (right) channels.

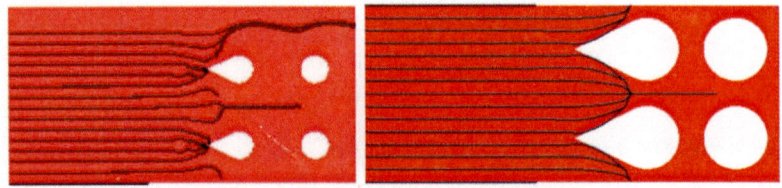

Figure 9.13. Particle trajectories corresponding to the trapping threshold for the HF (left) and Bosch (right) channels.

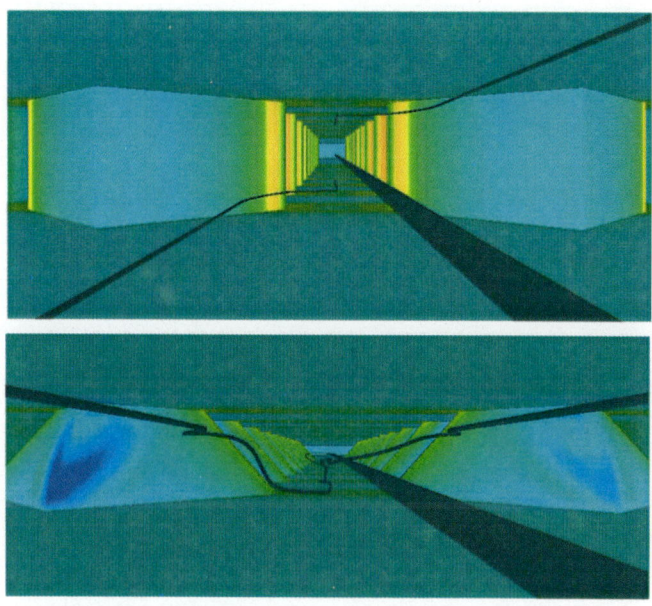

Figure 9.14. Particle trajectories, shown in black, as seen by particles within the channels, corresponding to the trapping threshold for the Bosch (upper) and HF (lower) channels.

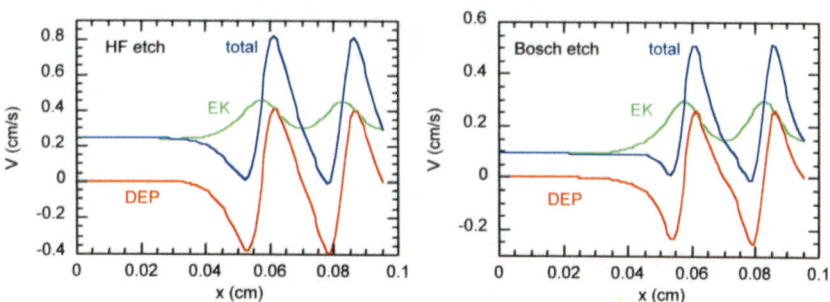

Figure 9.15. Total, EK and dielectrophoretic contributions to the particle velocities as functions of distance along the longitudinal centerline spanning the leading portion of the post array for the HF (left) and Bosch (right) channels.

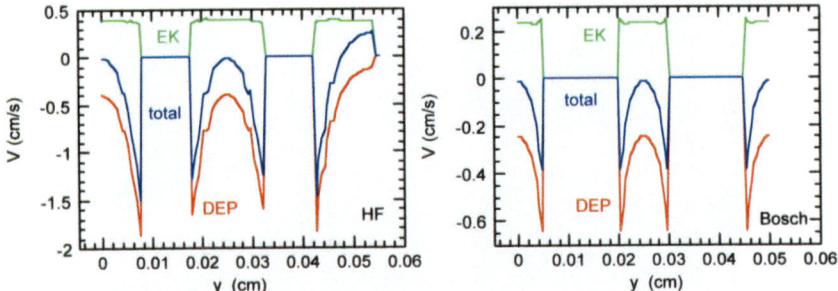

Figure 9.16. Total, EK and dielectrophoretic contributions to the particle velocities as functions of distance along the lateral line at the location of particle trapping at the first row of circular posts for the HF (left) and Bosch (right) channels.

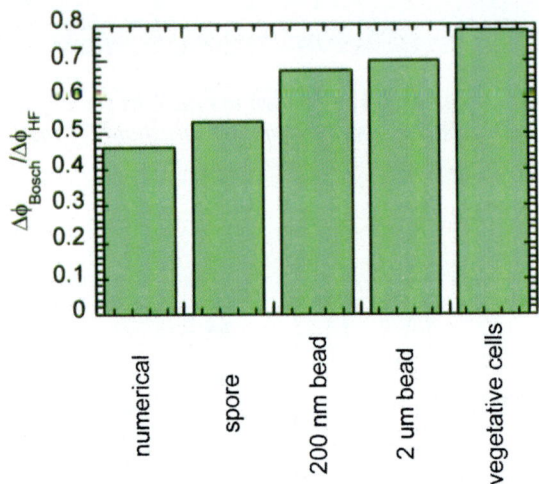

Figure 9.17. The ratio of the threshold trapping electric potential differences for the Bosch and HF channels as obtained from the calculations and as measured for several different types of particles.

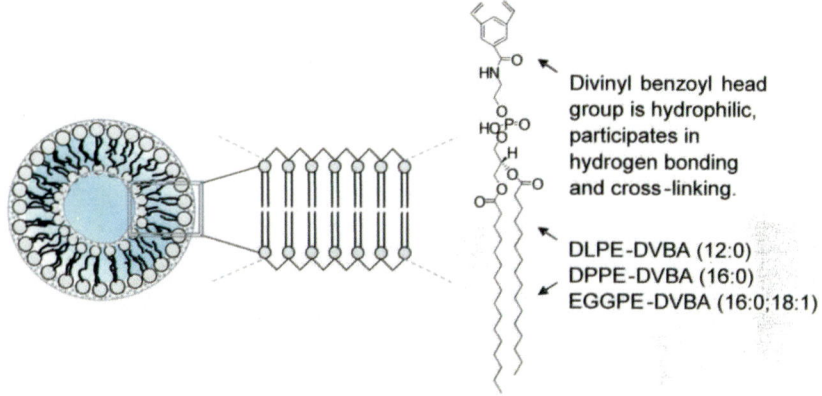

Figure 13.1. Schematic representation of cross-linked vesicle formed fromDLPE-DVBA, DPPE-DVBA or EGGPE-DVBA.

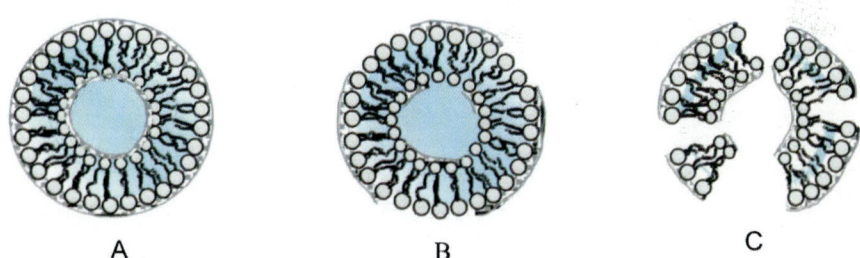

Figure 13.2. Graphic representation of vesicle cross-linking: (A) complete cross-linking; (B) partial cross-linking; (C) destructive cross-linking to vesicle structure.

Biological Detection

Chapter 9

A Comparison of Insulator-Based Dielectrophoretic Devices for the Monitoring and Separation of Waterborne Pathogens as a Function of Microfabrication Technique

Gregory J. McGraw, Michael Kanouff, Joseph T. Ceremuga,
Rafael V. Davalos, Blanca H. Lapizco-Encinas, Petra Mela,
Renee Shediac, John D. Brazzle, John T. Hachman,
Gregory J. Fiechtner, Eric B. Cummings, Yolanda Fintschenko,
and Blake A. Simmons

Sandia National Laboratories, 7011 East Avenue, Livermore, CA 94551

We present the selective trapping, concentration, and release of various biological organisms and inert beads utilizing insulator-based dielectrophoresis (iDEP) with discrete post arrays inside a polymeric microfluidic device. The arrays of insulating posts used to constrict electric field lines within the microchannel were fabricated using two of the most common fabrication techniques: chemical isotropic and reactive ion anisotropic etching. These fabrication methods, therefore, produce structures with distinct transverse profiles–the reactive ion etch produces relatively straight sidewalls, whereas the chemical etch produces tapered sidewalls. The devices were utilized to selectively separate and concentrate a variety of biological simulants and organisms. The dielectrophoretic responses of the organisms were observed to be a function of the applied electric field as well as post size and channel geometry. We then compare the device performance as a function of microfabrication methodology and through comparing experimental results and those predicted by computational modeling. We have found that there is a

© 2008 American Chemical Society

direct dependence on the cross-section profile of the channel and the non-uniform electric field gradients generated. This variation has a significant impact on the resultant performance and separation efficiency of the device.

Introduction

Insulator-based dielectrophoresis (iDEP), pioneered by Masuda et al. (*1*) and revisited by Lee *et al.* (*2*), employs insulating structures to produce the necessary spatial non-uniformities in an electric field. Devices for iDEP can be made purely from polymeric insulating materials (*3*). These devices can then be replicated in a very inexpensive commercial process, thereby facilitating high-throughput and large-volume applications. This paper demonstrates the effectiveness of such polymer-based devices. Chou et al. (*4*) has previously demonstrated iDEP of DNA molecules, *E. coli* cells, and red blood cells using insulating structures and AC electric fields. Recently, Zhou et al. (*5*) and Suehiro et al. (*6*) applied AC electric fields to channels packed with insulating glass beads to separate and concentrate yeast cells in water.

We have previously demonstrated both the theory and application of iDEP with D.C. electric fields using arrays of insulating posts inside a microchannel to trap polystyrene particles (*7*) and to separate live and dead bacteria (*8*). The effective separation between live and dead cells arose from differences between the membrane conductivities of the two classes of cells. When a cell dies, the cell membrane becomes permeable, and its conductivity can increase up to a value of 10 µS/cm; whereas the conductivity of the membrane of a live cell tends to be ~10^{-3} µS/cm. These differences in membrane conductivity dramatically change the Clausius-Mossotti factor, thereby producing significantly different dielectrophoretic trapping thresholds for the live and dead particles. While both exhibited negative dielectophoresis (DEP), the lower trapping threshold of the live cells allowed their selective collection, demonstrating the potential of iDEP for rapid cell viability analysis (*8*).

We have also reported the insulative dielectrophoretic separation between different species of live bacterial cells (*9*). In this case, parameters other than membrane conductivity play an important role in the separation process. These parameters include cell size, cell shape, and other morphological characteristics of the cells, such as the presence of a flagellum. While the theory is not yet complete enough to predict the relative trapping thresholds of different bacteria, we empirically showed that these thresholds are typically different enough between species of bacteria to allow selective collection. The threshold (minimum) applied electric field required to achieve dielectrophoretic trapping of the four species of bacteria in the study, from lowest to highest threshold, was *Escherichia coli* < *Bacillus megaterium* < *Bacillus subtilis* < *Bacillus cereus*

(9). These results demonstrate that iDEP can effectively and efficiently separate similar species of vegetative and viable bacterial cells.

Polymer microfluidic devices have been shown for several years in the fields of separation (10) and other lab-on-a-chip applications (11). Unlike other microsystem fabrication platforms, polymer devices can be cheaply and reliably mass produced since they are highly compatible with mass commercial fabrication techniques such as injection molding and hot embossing. This publication focuses on and presents the capabilities of polymer-based iDEP devices for the concentration and removal of water-borne bacteria, spores and inert particles using our micro-iDEP device. We demonstrate that the device performance can be directly linked to the cross-sectional transverse profile presented by the insulating structures. This is proven by fabricating the insulating structures with a taper and straight profile and observing device performance. The dielectrophoretic behavior exhibited by the different particles of interest (both biological and inert) in each of these systems was observed to be a function of both the applied electric field and the characteristics of the particle, such as size, shape, and conductivity. The results obtained illustrate the potential of polymer-based iDEP devices to act as a concentrator for a front-end device with significant homeland security applications for the threat analysis of bacteria, spores, and viruses. The polymeric devices exhibit the same iDEP behavior and efficacy in the field of use as their glass counterparts, but with the added benefit of being easily mass fabricated and developed in a variety of multi-scale formats that will allow for the realization of a truly high-throughput device. These results also demonstrate that the operating characteristics of the device can be tailored through the device fabrication technique utilized and the magnitude of the electric field gradient created within the insulating structures.

Experimental

Metrology

White light interferometry was employed to inspect the stamps and replicates used in this research in device fabrication. The Wyko NT3300 surface profiler is a commercially available inspection system equipped with hardware and software for data acquisition and analysis (www.veeco.com). The metrology system has 0.1 nm resolution in the vertical (z) axis. Laterally, in the XY plane, resolution is dependent on the pixel size in each scan. Thus, the objective and field of view of the scan determines lateral resolution. Scans in this study were performed using 20.8x effective magnification, translating to 0.7 μm pixel size. Considering the size scale of the structures involved, this lateral resolution was acceptable. Filters were applied to the scanned data sets to remove tilt so that the desired profilometry could be completed.

Tracers and Background Solutions

Inert particles

Carboxylate-modified polystyrene microspheres, *FluoSpheres*™, (Molecular Probes, Eugene, OR) having a density of 1.05 mg/mm^3 and diameters of 2 μm and 200 nm were utilized at a dilution of 1:10,000 from a 2% by wt stock suspension. Bead suspensions were sonicated between steps of serial dilution and before use.

Vegetative Cells

Bacillus subtilis (strain ATCC # 6633) were obtained from American Type Culture Collection (ATCC, Manassas, VA, USA). Cells were grown in 5 mL of Lennox L Broth (LB) at 30.7 °C, in an incubator, for 12 h, to achieve saturation conditions. A 1:20 volumetric dilution of the cell culture was then allowed to grow in the LB into late log phase to a cell concentration of 6 x 10^8 cells/mL, verified by optical density (OD) measurements at 600 nm. Cells were centrifuged at 5000 rpm for 10 min, in order to eliminate the LB, and resuspended in deionized (DI) water (pH 8) utilizing a vortex mixer. The cells were then labeled with Syto®17 bacterial stain (Molecular Probes, Eugene, OR, USA). Syto®17 produces cells that exhibit red fluorescence (excitation/emission 621/634 nm). For every milliliter of cell culture present in the vial, 9 μL of the fluorescent nucleic acid stain were added. The cells were then incubated at room temperature for 15 min. The labeled cells were recovered by centrifugation at 5000 rpm for 10 min, washed three times with DI water to remove any free dye, and finally resuspended in DI water to the desired final volume to reach the desired cell concentration (typically 6 x 10^8 cells/mL).

Spores

Spore suspensions of *B. subtilis* (strain ATCC # 6633) and *Bacillus thuringiensis* (strain ATCC # 29730) were obtained from Raven Biological Laboratories Inc. (Omaha, NE). The spore samples were labeled with Syto® 11 (Molecular Probes, Eugene, OR) dyes that labels the spores fluorescent green (excitation/emission 508/527 nm). The samples were dyed as received, without any further modification. The spore samples were labeled by following the same protocol used with the vegetative bacterial cells. The final concentration of the labeled spores was 1 x 10^9 spores/mL.

Background solution

Deionized water from a reverse osmosis filter (Millipore) was titrated with KOH and HCl to a pH of approximately 8. Conductivity was then adjusted by titration with KCl to an endpoint of 4 μS/cm.

Device Fabrication

Microfluidic channels were arranged on bonded discs of Zeonor® 1060R resin (Zeon Chemicals, Tokyo, Japan). The lower discs were injection molded, using a custom mold with a negative of the microchannel troughs and posts on its surface. The mold was fabricated using a glass or silicon master, with the microchannel features photolithographically etched onto its surface. Since the mold is a negative of the master, the master and the final polymer disc contain the same features. The microchannels are arranged in 2 rows of 4 on each of the masters. Each microchannel was 10.2 mm in length and 1 mm wide. A rectangular array of insulating posts is placed in the center of the channel, 2.9 mm from each end. Each disc featured two different array patterns, a 4 x 10 post array and a 5 x 12 post array. Only channels featuring the 4 x 10 array were used for the present DEP experiments. The 4 x 10 array was composed circular posts with 200 μm diameters spaced 250 μm center-to-center. The rows at the front and back of the array featured a pointed surface facing outward to reduce fouling. Two distinct masters were produced using different standard etch techniques. In the first, a photolithographic pattern was defined on Schott D263 glass wafers and etched with hydrofluoric (HF) acid. This produced a stamp with characteristic height of 55 μm. The other fabrication process, a deep reactive ion etch (Bosch), was created by utilizing a different mask material patterned on a silicon wafer and produced 75 μm features.

After patterning, the masters were sputter coated with an electroplating base material, in this case 1500 Å of chrome (glass) or titanium (silicon) for adhesion promotion and 1500 Å of copper. The masters were then placed into a Digital Matrix commercial DM3M electroplating machine. The bath chemistry utilized was a standard nickel sulfamate with controlled pH to minimize stresses. Electroplating occurred at 48 °C for a total of 40 amp-hours and produced nickel films with thicknesses typically on the order of 1 mm. The nickel was then planarized and machined to the set dimensions for use in our custom, in-house, fabrication facilities and the glass/silicon and metal seed layers dissolved. The nickel stamp was then thoroughly characterized through metrology, visual inspection, and electron microscopy.

Injection molding was carried out utilizing a 60-ton Nissei® (Nissei® America, Los Angeles, CA) TH-60 vertical injection molding machine. Pellets of Zeonor® 1060R resin were dried at 40 °C for at least 24 hr before use. The resin was then fed to the machine through a gravity-assisted hopper connected externally to the injection molding barrel. Injection molding conditions were empirically determined by the operators, using the polymer supplier's recommendations as a starting point for molding. Cross-polarized optical interrogation of the replicated substrates was employed to assess and minimize residual stresses in the injection molded parts.

Manufactured discs of Zeonor® 1060R from Zeon Chemicals (Louisville, KY) with a planar surface and a thickness of 1.5 mm were used as the upper disc to seal the channel. Through ports ("vias") of 1 mm diameter were drilled through the upper disc to provide a fluidic interface and reservoir at each end of the 8 microfluidic channels on the lower disc. The upper and lower discs were then thermally bonded using a Carver (Carver, Inc., Wabash, IN) press. Bonding conditions were held constant at the following: the press was heated to 190 °F with a constant applied load of 750 psi and a corresponding cycle time at temperature of 60 min. The bonded assembly was then cooled to 75 °F under constant load and then removed from the press. All bonded assemblies were checked for flow and channel blockage before use.

The bonded disc was reversibly sealed to the base of a custom PDMS manifold using a vacuum chuck. The manifold is ported with 16 openings spanning its thickness that coincide with the inlet and outlet vias of each channel. Each opening can accept a slip tip syringe and forms a watertight seal between the syringe and the drilled via. The channels were primed by gently forcing background solution through the channel with a syringe. Bubbles were removed with suction, as necessary. After priming with DI water, 3-mL slip tip syringes, with plungers removed, were loaded with approximately 1.5 mL of the desired tracer and inserted at the upstream and downstream ports of the manifold. The tracer was then gently forced into the channel with a plunger until the expected concentration of particles could be visualized. 0.508-mm diameter platinum-wire electrodes (Omega Engineering Inc., Stanford, CT) were inserted directly into the syringes. The positive electrode always corresponds to the upstream direction, since Zeonor® 1060R has a negative zeta potential at pH ≈ 8. A programmable high voltage sequencer, Labsmith HVS 448 (Livermore, CA) was used to apply voltages of 1500 V and below. A manually-controlled power supply, Bertran ARB 30 (Valhalla, NY) was used for higher voltages. The apparatus was visualized with an inverted epifluorescence microscope, model IX-70 (Olympus, Napa, CA). Different sets of fluorescence filters were employed: Chroma 51006, Chroma 51004 (Chroma Technologies Corp, Brattleboro, VT) and Olympus 41012 (Olympus, Napa, CA).

Determination of Trapping Thresholds

Suspensions of each of the four tracers were prepared. Biological agents were diluted approximately 1:20 from the stock suspensions and vortex mixed. Once the syringes were loaded with tracer and the electrodes were inserted, the difference in hydrostatic pressure between syringes was balanced by adjusting the level of background solution in the downstream syringe until flow was observed to stop.

The trapping threshold is defined as the minimum voltage for which trapping of particles is observed near multiple posts within 3 sec of the initial application of voltage and for which trapping is sustained for at least 10 sec. This was determined by a binary survey of discrete voltages, beginning with trapping voltages observed for glass chips (9). After each test voltage is applied, approximately 1 mL of liquid is removed from the downstream syringe and fresh suspension is allowed to flow through the channel under the resulting pressure head. Once several channel lengths of tracer solution have fed through the channel, the liquid level in the syringes was re-balanced to eliminate pressure driven flow. Then, the next voltage is tested. The endpoint of the survey was reached when a minimum trapping voltage was identified within 25 V_{dc} of a non-trapping voltage.

Once the first trapping threshold voltage is determined, at least two more trapping threshold voltages were then obtained for each tracer type. The search for each subsequent voltage began at 100 V_{dc} less than the trapping threshold determined in the previous trial. Trapping threshold measurements for each tracer type were performed in the same channel, unless otherwise noted. Measurements for different types of tracers were performed in separate channels.

Safety Considerations

The use of high voltage is a hazard that requires training and safety measures, such as an interlocks and current-limiting features. The Syto® series labels were handled with care. All organisms used are classified as risk group 1 organisms by U. S. Centers for Disease Control and Prevention (CDC) and ATCC and were handled in a BSL1-approved facility. Care was taken to handle BSL1 materials and dispose of the waste according to the CDC guidelines and the policies of Sandia National Laboratories.

Modeling Protocols

Calculations were carried out for the relative trapping performances of the Bosch and HF channels. This required the calculation of the electric fields in each of the channels and using them to calculate trajectories of particles driven by a combination of electroosmosis, electrophoresis, and DEP.

The liquids of interest here are electrically neutral, except for the Debye layers next to the channel walls. Since Debye layers on microchannel walls are usually thin (10 nm) compared to lateral channel dimensions (50 mm), the electric potential, ϕ, satisfies the Laplace equation. In terms of the dimensionless electric potential, $\phi^* = \phi/\Delta\phi$, where $\Delta\phi$ is the potential difference applied across the channel, this is given as Equation 1,

$$\nabla^2 \phi^* = 0, \qquad (1)$$

where the operator is assumed to be dimensionless. The dimensionless electric field, $E^* = EL/\Delta\phi$, is obtained from the electric potential as $E^* = \nabla\phi^*$, where E is the dimensional electric field and L is the length of the channel. Equation 1 shows that, for a given geometry, there is only one solution for ϕ^*, i.e., there are no parameters. Solutions for the dimensional variables, ϕ and E, for specific values of $\Delta\phi$ and L can be obtained from the lone solution for ϕ^* by simply multiplying it, and its gradient, by $\Delta\phi$ and $\Delta\phi/L$, respectively.

Particle trajectories are determined by a combination of fluid-flow, electrophoresis and DEP. The fluid-flow is driven by electroosmosis for the case of interest here. For the thin Debye layer approximation, electroosmotic flow may be simply modeled with a slip velocity adjacent to the channel walls that is proportional to the tangential component of the local electric field, as shown by the Helmholtz-Smoluchowski equation (12). Here, the proportionality constant between the velocity and field is called the electroosmotic mobility, μ_{EO}. Fluid-flow in microchannels becomes even simpler for ideal flow conditions where the zeta potential, and hence μ_{EO}, is uniform over all walls and where there are no pressure gradients. For these conditions, it can be shown that the fluid velocity at all points in the fluid domain is given by the product of the local electric field and μ_{EO} (13).

In general, particle velocities can differ from fluid velocities due to electrophoretic and dielectrophoretic forces. The particle velocity induced by electrophoresis, relative to the fluid velocity, is given by the product of the electric field and the electrophoretic mobility, μ_{EP}. Similarly, the dielectrophoretic-induced particle velocity is given by the product of the gradient of the electric field squared, $\nabla(E \cdot E)$, with the dielectrophoretic mobility, μ_{DEP}. The total particle velocity is given by the sum of all of these terms as shown by Equation 2,

$$v_{particle} = \mu_{EK} E + \mu_{DEP} \nabla(E \cdot E), \qquad (2)$$

where the EK mobility is defined as, $\mu_{EK} = \mu_{EP} + \mu_{EO}$. The mobilities, μ_{DEP} and μ_{EP}, depend on characteristics of the particle and surrounding fluid. Here, we are interested in predicting the ratio of threshold electric potentials, $\Delta\phi_{th}$, that must be applied to the Bosch and HF channels in order to trap various particles. For this, it is not necessary to find values for the mobilities, as described below.

A particle is trapped when it is driven into a region where the dielectrophoretic and EK forces balance and the particle velocity goes to zero. Equation 2 shows that this balance depends on E, μ_{DEP} and μ_{EK}. Scaling Equation 2 by $\mu_{EK}\Delta\phi/L$ such that it becomes dimensionless yields Equation 3,

$$v^*_{particle} = E^* + S\nabla(E^* \cdot E^*), \qquad (3)$$

where $S = \Delta\phi\mu_{DEP}/(L^2\mu_{EK})$, which is a dimensionless measure of the relative importance of DEP and electrokinesis. Low values of S result in insufficient dielectrophoretic forces to trap particles. There is a threshold value for S, S_{th}, that results in a dielectrophoretic force just large enough to counter the EK forces. The corresponding threshold value of $\Delta\phi$ can be obtained from S_{th} as $\Delta\phi_{th} = (L^2\mu_{EK}/\mu_{DEP})S_{th}$. Here, we are interested in the ratio of values of $\Delta\phi_{th}$ for the Bosch and HF channels, which is given by $\Delta\phi_{th\text{-}Bosch}/\Delta\phi_{th\text{-}HF} = S_{th\text{-}Bosch}/S_{th\text{-}HF}$, since L, μ_{EK}, and μ_{DEP} are the same for both channels for a given particle type. Particle trajectory calculations were carried out to identify values for S_{th} based on Equation 3, using the solutions for E^* obtained for the Bosch-etched and HF-etched geometries. Note that only particles that experience negative DEP are considered here, i.e., $\mu_{DEP} < 0$, where the dielectrophoretic force acts to drive particles away from regions of large field intensity.

Results and Discussion

Metrology

Dimensional characterization gives insight to how well the microfluidic structures conform to design specifications. There are a number of reasons to perform dimensional metrology on the stamps and replicates used in this study. The stamp/replicate geometry determines fluid-flow characteristics, so understanding the geometry aids in diagnosing performance problems that may arise due to distortions or defects. A quality assessment of the replication process can be made by comparing data from replicates to their corresponding stamp. A comparison between the two etch processes used to pattern the stamp may also be concluded. Moreover, the gathered metrology data provides a baseline for process control and improvement.

The specimens are fabricated using an injection molding process in which the structures are formed on metal stamps using etching processes. The sidewall normality depends on the etch process used. The glass (HF) etching process produces structures that have tapered/sloped sidewall geometries (*14*). The Bosch etch creates more of a vertical sidewall; however there are corrugations in the hoop direction from bottom to top of the structures (*15*). The size scale of these features is on the order of tens of microns to millimeters laterally and tens to hundreds of microns in height. Dimensional metrology of structures having these characteristic size and topography is challenging. Non-contact profilometry, using a white-light interferometer and the technique utilized, is limited because of signal loss once the topography of the structure exceeds a certain slope. Contact profilometry was avoided because the probe tip must be smaller than the feature size. Also, the stylus must have the necessary clearance to accommodate the slope of the structure's sidewall to avoid interference.

Various measurements were taken to characterize the replication process: post height, surface roughness, vertical range (Top Δz) on the circular post, vertical range (Bottom Δz) on the surrounding surface, and surface area of the circular post. Figure 1 shows a principal sketch of the dimensions measured.

Data acquired by the interferometer is meaningful, despite the signal loss in areas of sloped topography, because one goal of this metrology is to compare the stamp to the replicate. If the replication process was ideal, the signal loss seen when inspecting the stamp would transfer over to the plastic replicate. This is the reason for calculating the surface area of the posts. To characterize the replication process, data was taken from an HF glass-etched stamp and two of its replicates and a Bosch-etched stamp and three of its replicates. Each stamp had similar layouts, consisting of circular posts in rectangular arrays.

The post heights were 55 μm and 75 μm for the wet HF and Bosch etches, respectively. Three posts were chosen on each sample: two 200-μm diameter

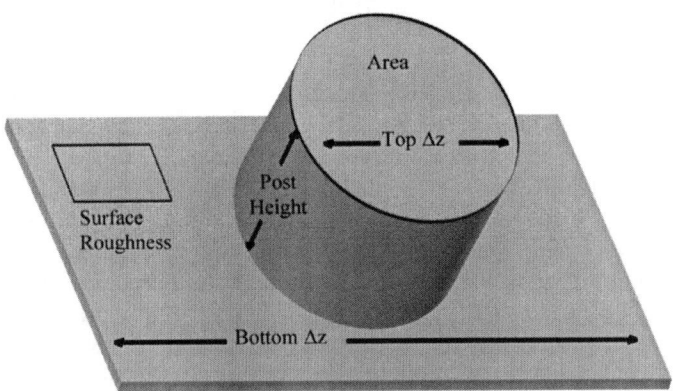

Figure 1. Schematic depiction of metrology measurements of the post structures utilized for comparison. (See page 1 of color inserts.)

posts on the 4 x 10 post pattern and one 150-μm diameter post on the 5 x 12 post pattern. To guarantee consistency, one objective and field of view setting was chosen based on maximizing the feature size within the field of view across all features inspected. Corresponding posts from stamp to replicate were examined. Figure 2 gives a qualitative perspective on the stamp to replicate quality, as well as the repeatability of the replication from both stamp types.

Table I contains data referring to the nominal design specification, the stamp, the replicate, and the difference between the replicate and the stamp, in order to quantify the replication process. The values in Table I are expressed as averages across the multiple samples acquired.

Considering the data presented in Table I, it is noted that the characteristic post height feature measured replicates to a deviation less than 2 μm from the stamp, indicating good feature fidelity and minimal underfill during the injection molding process. The bottom surfaces of the plastic replicates have minimal differences in height variation and average surface roughness as well, reinforcing this conclusion.

The top surfaces of the replicate posts show small deviations from the stamp as well. There are some observed distortions in the topography of the replicate posts. These distortions are due to the adhesion of the plastic to the metal during release of the replicate from the stamp. The performance of the post pattern is not compromised, because fluids are intended to move around the posts and not over the top surface. The area measurements from nominal-to-stamp and from stamp-to-replicate for the Bosch etch have a higher overall deviations from the nominal structures when directly compared to the HF glass etch post structures.

Table I. Replication process measurements

Bosch Etch	Post Height (μm)	Bottom Δz (μm)	Bottom R_a (nm)	Top Δz (μm)	Area^a (mm^2)	Area^b (mm^2)
Nominal	75.0	0.0	na	0.0	0.0314	0.0177
Stamp	74.0	1.9	61.1	11.5	0.0320	0.0180
Replicate	74.4	4.3	57.3	12.4	0.0250	0.0097
Replication Δ	0.4	2.4	-3.9	0.9	-0.0070	-0.0083
HF Glass Etch	Post Height (μm)	Bottom Δz (μm)	Bottom R_a (nm)	Top Δz (μm)	Area^a (mm^2)	Area^b (mm^2)
Nominal	55.0	0.0	na	0.0	0.0310	0.0177
Stamp	52.8	2.1	34.6	2.6	0.0055	0.0020
Replicate	51.0	2.0	31.6	2.9	0.0063	0.0002
Replication Δ	-1.8	-0.1	-3.0	0.3	0.0008	-0.0018

[a] Area measurement for 200-μm diameter posts.
[b] Area measurement for 150-μm diameter posts.

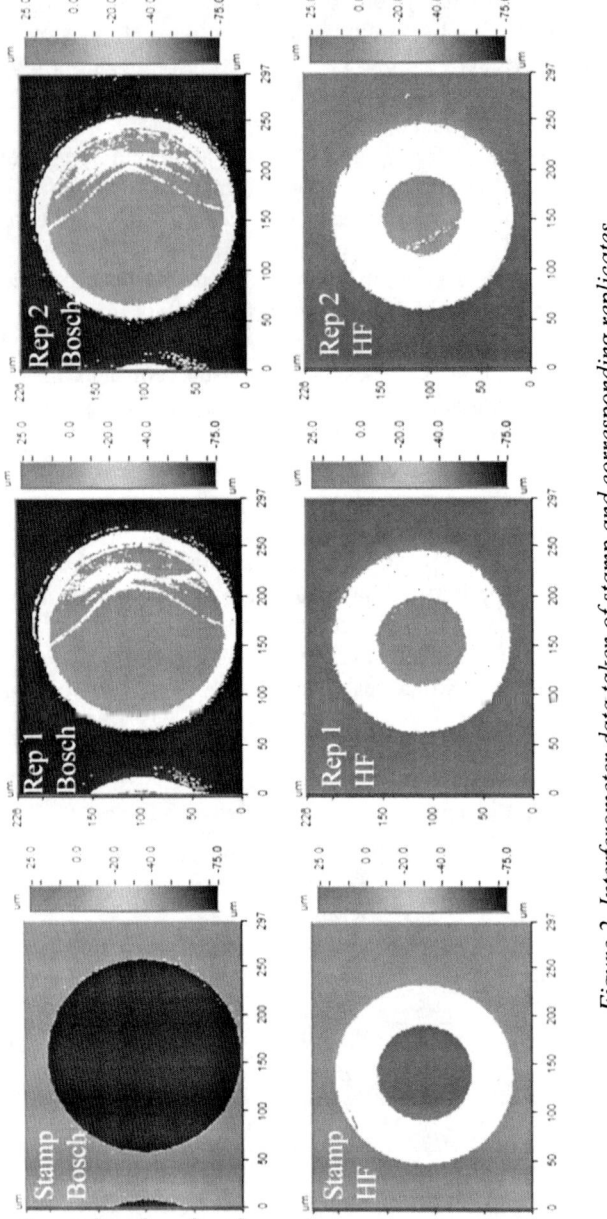

Figure 2. Interferometer data taken of stamp and corresponding replicates for 200-μm diameter posts for Bosch and wet HF replication. (See page 2 of color inserts.)

iDEP Device Performance Characterization

After the devices were characterized with metrology and sealed to form watertight channels, the next step was to evaluate the performance of the devices as a function of fabrication technique. The experimental apparatus utilized for all of the experiments is shown in Figure 3.

Trapping thresholds were clearly determined by direct observation of particle behavior through epifluorescence microscopy of the microchannels. At a discrete trapping voltage, particles begin to collect in spaces between posts, as seen in Figure 4a. This process was found to be highly reversible, with concentrated streams of particles leaving the posts when voltage is discontinued, as shown in Figure 4b. Differential separations based on type of particle were studied in the microchannels formed from the Bosch etch mold. Since the local maxima of the electric field are located in the narrowest region of the gap between posts, particles with a lower trapping threshold will be stopped further away from this region than particles with higher trapping thresholds. This creates the differential banding observed in Figures 5b, 6b, and 6c.

At voltages intermediate between the trapping thresholds of two tracers, the less readily trapped tracer passed through, while the more readily trapped tracer was retained in the posts. It was observed that 2-μm polystyrene beads were separated from 1-μm polystyrene beads, as shown in Figure 5.

Different types of biological tracers were also separated from each other and from non-biological backgrounds and presented in Figure 6. *B. subtilis* spores were separated from a mixture of spores and 1-μm polystyrene beads. Additionally, to show device operation and selectivity in the presence of an inert background, *B. subtilis* spores were separated from a dense background of 20-nm polystyrene beads. *B. subtilis* spores and vegetative cells formed separate bands of trapping at 170 V/mm (Figure 6a). Finally, *B. subtilis* and *B. thuringiensis* spores were separated into bands at 200 V/mm (Figure 6b). Surprisingly, the trapped *B. thuringiensis* formed the inner band in this separation. *B. thuringiensis* spores are slightly larger, on average, than *B. subtilis* spores, so one would expect them to form the outer band based on size alone. This emphasizes the importance of factors such as particle composition and morphology in determining trapping threshold.

The trapping thresholds for the individual tracer types were found to scale roughly as expected, with particle trapping threshold decreasing with increasing particle size. A major exception to this trend is that biological particles appear to trap more easily than polystyrene beads of comparable size. Vegetative *B. subtilis* cells trapped at a lower threshold voltage than 2-μm polystyrene beads, despite having a smaller internal volume than the beads. Otherwise, the order of trapping threshold by increasing voltage was 2-μm beads, *B. subtilis* spores, and 200-nm beads, which corresponds to the relative tracer sizes. In a qualitative study, *B. subtilis* spores were observed to trap more readily than 1-μm polystyrene beads, despite having a smaller average size than the 1-μm beads.

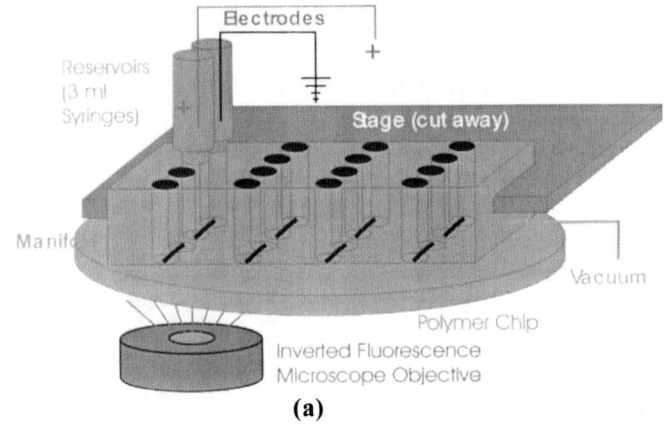

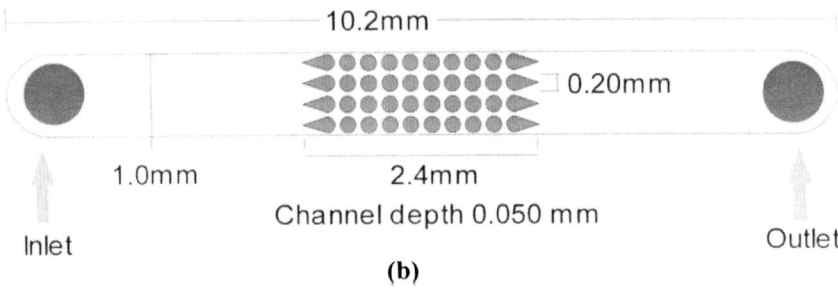

*Figure 3. Schematic depiction of the experimental setup showing (a) each device contained 8 individual channels on a chip injection molded from Zeonor® 1060R resin. The chip was sealed to a manifold with a vacuum chuck and the apparatus was clamped to a holder on top of the microscope stage. (b) A schematic of a channel with 4 x 10 post configuration.
(See page 3 of color inserts.)*

This is likely due to the high resistivity of the spore coat, when compared with the resistivity of the polystyrene beads. A summary of the trapping thresholds determined for both device types is presented in Figure 7.

The trends in the experimentally observed trapping thresholds agree well with expectation based on previous work in glass microchannels (9). Particles trap at lower voltages in microchannels molded with the stamp made from the Bosch-etch master. Conversely, particles require higher voltages to trap in microchannels molded with the stamp made from the HF-etch master. The ratio of the trapping voltages for the Bosch and wet etch cases is between 0.54 and 0.79. This may be a function of particle composition or shape, since the ratios observed for the 2-µm and 200-nm spheres are nearly identical.

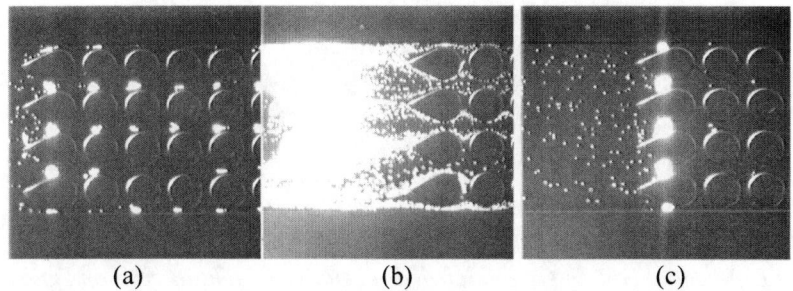

Figure 4. Still photographs, taken from a digital movie, demonstrating trapping of 2-μm polystyrene beads in an array molded from the Bosch etch-derived device. Near the trapping threshold (a), array behaves analogously to a packed-bed adsorber. Upstream posts saturate and pass particles to traps downstream, creating a saturation wavefront at startup. When the trapping voltage is turned off (b), particles flow out of traps in a concentrated plug. At higher trapping voltages (c), trapped particles remain in the forward posts.

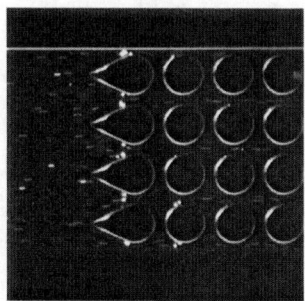

Figure 5. Demonstration of separation of polystyrene beads by size. 2-μm red beads trap at 50 V/mm, while 1-μm green beads pass between posts. (See page 3 of color inserts.)

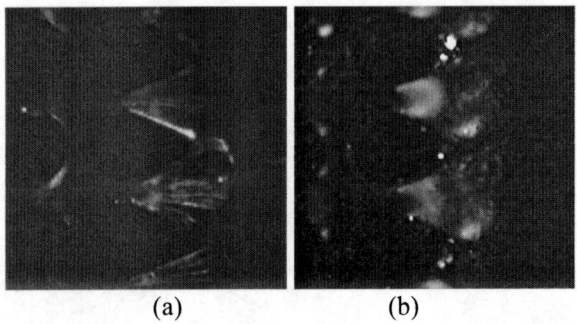

Figure 6. Differential trapping comparison of biological tracers. (a) Differential banding of red-labeled vegetative B. subtilis cells and green-labeled B. subtilis spores at 170 V/mm. (b) Differential banding of green-labeled B. subtilis spores and red-labeled B. thuringiensis spores at 200 V/mm. (See page 4 of color inserts.)

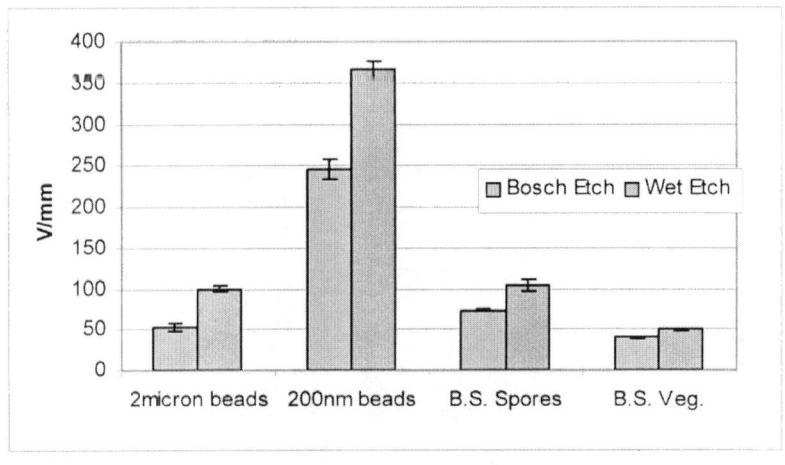

Figure 7. Trapping thresholds of particles studied as a function of device fabrication technique. The particle trapping threshold is observed to be a function of particle size, particle conductivity, and morphology. (See page 4 of color inserts.)

Computational Modeling Comparison of Device Performance

After the trapping thresholds of the device were determined experimentally, we proceeded to evaluate the device performance based on rigorous computational modeling algorithms, as a means of comparison of real and expected trapping threshold values. Due to the complex geometries considered here, Equation 1 is solved using the finite element method. The finite element meshes for the HF and Bosch channels are shown in Figure 8, where only half of each channel is included due to symmetry. The boundary conditions are, $\phi^* = 1$ at $x/L = x^* = 0$, and $\phi^* = 0$ at $x^* = 1$, and $\nabla \phi^* \cdot n = 0$ over all other surfaces, where n is the local unit vector normal to the surface. The solution for ϕ^* must be differentiated twice to get the dielectrophoretic force on a particle. Consequently, in order to obtain smooth results for this force, it was necessary to use a very fine mesh. The meshes shown in Figure 8 contain over one million elements.

Figure 8. The finite element meshes of the HF (upper) and Bosch (lower) iDEP channels. (See page 5 of color inserts.)

Figure 9 shows the solution for the electric potential along the longitudinal centerline. There are three distinct regions in the solution, the post array region, the open region of the channel leading up to the post array region, and another open region leading away from the array region.

The gradient in ϕ^* is the largest within the array region due to the restrictions in channel cross section provided by the posts. The gradient is larger for the Bosch channel than the HF channel because of the vertical side walls in the former, which results in a smaller open cross section. This results in larger

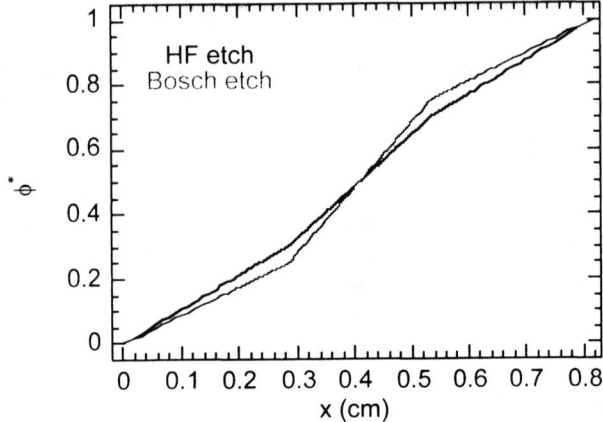

Figure 9. The non-dimensional electric potential as a function of distance along the longitudinal centerlines of the Bosch and HF channels. (See page 5 of color inserts.)

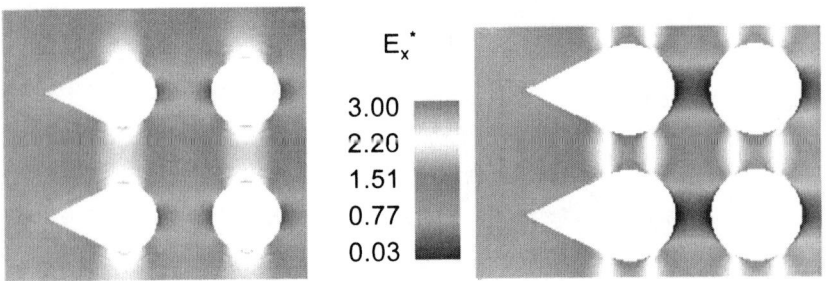

Figure 10. The x-component of the non-dimensional electric field in the leading region of the post array for the HF (left) and Bosch (right) channels. (See page 6 of color inserts.)

values for E^* and $\nabla(E^* \cdot E^*)$, and is the primary cause of differences in trapping performance between the two channels. Figure 10 shows the x-component of the non-dimensional electric field, E^*, in the leading region of the post array. The larger field strength between the posts in the Bosch channel can be seen clearly.

The x-component of the gradient of the electric field intensity, $\nabla(E^* \cdot E^*)_x$, is shown in Figure 11. Larger values exist between the posts for the Bosch channel compared to the HF channel due to the tighter restrictions between channels in the former, as discussed above for the electric field.

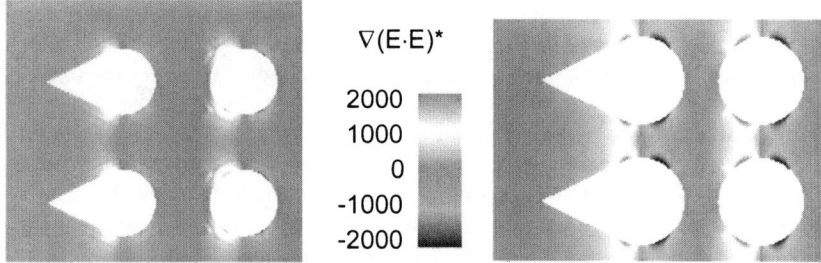

Figure 11. The x-component of the non-dimensional gradient of the electric field intensity in the leading region of the post array for the HF (left) and Bosch (right) channels. (See page 6 of color inserts.)

Figure 12 shows the distributions of E^*_x and $\nabla(E^* \cdot E^*)_x$ along a portion of the longitudinal centerline line spanning the leading region of the post array for the Bosch and HF channels. Both E^*_x and $\nabla(E^* \cdot E^*)_x$ are larger between the posts for the Bosch channel than the HF channel. The first peak value of $\nabla(E^* \cdot E^*)_x$ is slightly smaller than the second peak value, corresponding to the teardrop shaped posts and the first row of circular posts, respectively.

Particle trajectory calculations were carried out using Equation 3 and the solutions for the electric field presented above. Values of S_{th} were found for each channel that corresponded to their trapping thresholds. Particles were evenly distributed over the inlet to the channels to start the calculations. Results for

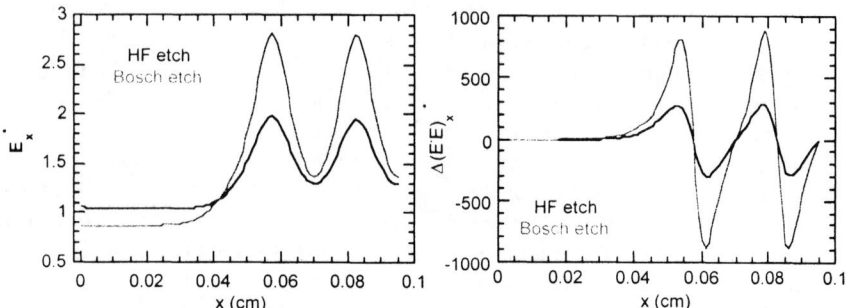

Figure 12. The x-component of the non-dimensional electric field (left) and gradient of the electric field intensity (right) as functions of distance along the longitudinal centerline spanning the leading region of the post array for the HF and Bosch channels. (See page 5 of color inserts.)

particle trajectories corresponding to the trapping threshold are shown in Figure 13. The particles are driven away from the posts, for the most part, and are trapped between the posts. Most of the particles are able to pass the "teardrop" shaped posts and are trapped at the first row of circular posts where the dielectrophoretic force is larger. It should be noted that the particle flow around the HF posts is indeed symmetrical, but is truncated along the bottom portion of the figure as presented. Particles are able to completely pass the post array in the HF channel along the outer region of the post array due to the relatively wide open path available there. These particles were not included when determining trapping threshold.

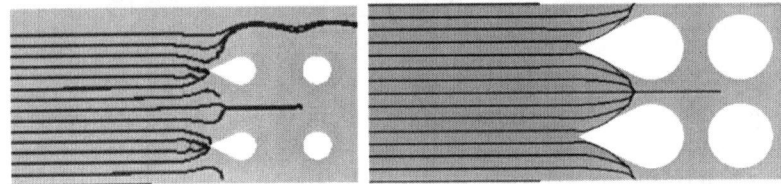

Figure 13. Particle trajectories corresponding to the trapping threshold for the HF (left) and Bosch (right) channels. (See page 6 of color inserts.)

Figure 14 shows a transverse channel perspective of several particle trajectories within the channels. This figure gives a clear view of the differences in geometry of the two channels. Estimates were made for the mobility values for the particles based on measurements of the EK flow velocity observed in regions of the channel remote from the post array (μ_{EK} = 0.00023 cm^2/s•V), and the threshold trapping voltage measured for spores (μ_{DEP} = 1.38 • 10^{-9} cm^4/s•V^2). These were used to calculate dimensional values for the particle velocity, including the individual contributions from electrokinesis and DEP. These are shown in Figure 15 as functions of distance along the longitudinal centerline spanning the leading portion of the post array, and in Figure 16 as functions of distance along a lateral line at the location of particle trapping at the first row of circular posts. Due to the smaller value of $\nabla(E^* \cdot E^*)_x$ for the HF channel, a larger electric potential difference is required to trap the particles than for the Bosch channel. The EK and DEP contributions to the particle velocities are different for the Bosch and HF channels, but at the location where the particles are trapped, the two contributions add to zero (the total velocity) for both channels. The plot on the right side of Figure 15 shows the lateral distribution of the particle velocities. This shows that the total velocity is negative everywhere at the longitudinal location of the trapped particles, except at the midpoint between the posts where it is zero. Thus, the weakest part of the trap is at this midpoint.

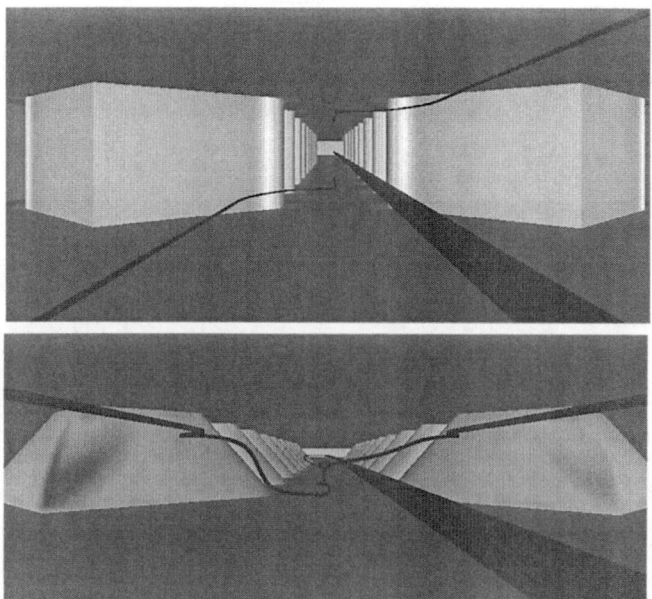

Figure 14. Particle trajectories, shown in black, as seen by particles within the channels, corresponding to the trapping threshold for the Bosch (upper) and HF (lower) channels. (See page 7 of color inserts.)

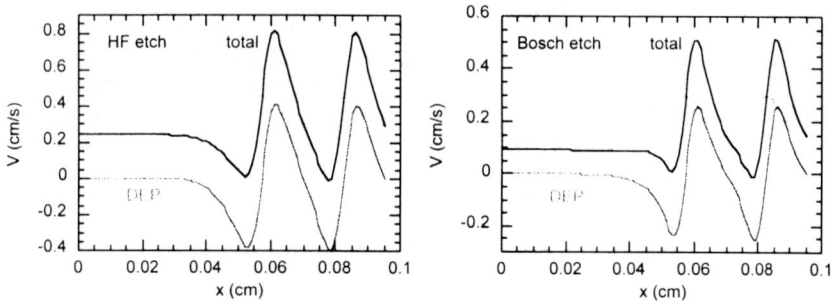

Figure 15. Total, EK and dielectrophoretic contributions to the particle velocities as functions of distance along the longitudinal centerline spanning the leading portion of the post array for the HF (left) and Bosch (right) channels. (See page 7 of color inserts.)

Figure 17 shows the ratio of the threshold trapping electric potential differences for the Bosch and HF channels obtained from the calculations along with the experimental values obtained for a number of different particle types. The calculated value of 0.46 is in reasonable agreement with that measured for spores, but the values for the other types of particles are much larger.

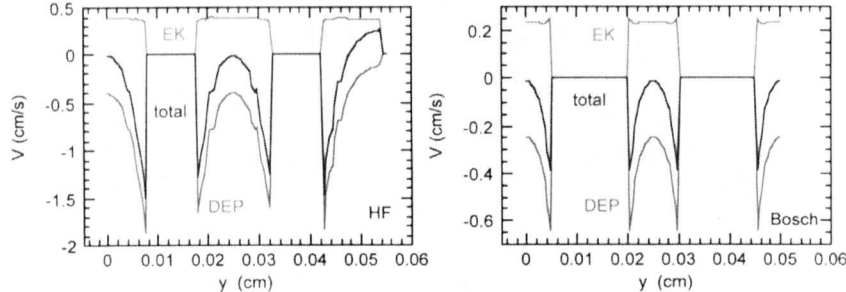

Figure 16. Total, EK and dielectrophoretic contributions to the particle velocities as functions of distance along the lateral line at the location of particle trapping at the first row of circular posts for the HF (left) and Bosch (right) channels. (See page 8 of color inserts.)

One possible explanation for this difference between the experimental and theoretical values for the ratio of threshold trapping potentials is the presence of particle-particle interactions that may perturb the electric field gradient present between the posts. The local perturbation to the field created by one particle may amplify the dielectrophoretic force acting on a nearby particle. A large number of such interactions may cascade, resulting in a significant enhancement of the dielectrophoretic force. This could result in direct contact between particles and van der Waals forces would then act to keep the particle ensemble together.

Summary

Microfluidic devices fabricated from a polymer substrate have been demonstrated to be effective at trapping and concentrating suspended organic particles in flowing water using insulative DEP. We have presented here a direct structure-function relationship between the taper of the insulating structures that is dictated by the microfabrication technique. In the present work, microfluidic channels were injection molded from Zeonor® 1060R using two different molds based on the same mask pattern. One mold was

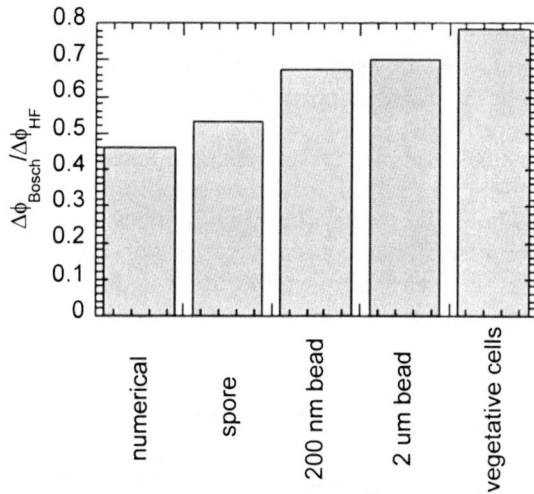

Figure 17. The ratio of the threshold trapping electric potential differences for the Bosch and HF channels as obtained from the calculations and as measured for several different types of particles.
(See page 8 of color inserts.)

electroplated onto an isotropically etched master, while the other mold was electroplated onto a master produced with an anisotropic etch process. The isotropic etch (HF) produced features with a vertical taper, while the anisotropic etch (Bosch) produced features with straight sidewalls. The topography of the resulting devices was thoroughly characterized with metrology and it was found that injection molding reproduced the features of the original masters with a high degree of fidelity.

Sealed microchannels were loaded with different particle suspensions. The minimum DC voltage required to separate each particle type from the surrounding fluid was determined. Significantly higher voltages were required to separate particles in microchannels with tapered (HF) features than in microchannels with straight-walled (Bosch) features. Differential separation of multiple particle types was qualitatively demonstrated. Generally, smaller particles required higher voltage to separate and microorganisms required less voltage to trap than similarly sized polystyrene microspheres.

The metrology data was used to create a numerical model for the dielectrophoretic devices. The calculated electric field profiles illustrate that non-tapered features are much more effective at creating local variations in field intensity than tapered features. The dynamics of each channel type were studied by the introduction of virtual tracers into the calculated electric field and a ratio of expected threshold voltages required to trap a particle in each channel type was calculated. While the model explained many of the experimentally observed performance differences between the two different microchannel geometries, there were significant differences between the experimentally observed threshold voltage ratios and the calculated ratio. Both the experimental and modeling studies of iDEP in polymer microchannels strongly suggest that iDEP can be an efficient, selective, and cost effective method for pre-concentrating biological samples for analysis in a microfluidic system. While there remains some disagreement between the theoretical and experimental results presented, the present work clearly illustrates the potential utility of an iDEP device that can isolate and concentrate targeted particles of interest from a diverse background. Such devices may prove useful in the isolation and detection of biological agents for homeland security applications.

Acknowledgments

The authors gratefully acknowledge the assistance of Judith Rognlien, Blanca Lapizco-Encinas, and Allen J. Salmi over the course of this project. This work was performed by Sandia National Laboratories for the United States Department of Energy under Contract DE-AC04-04AL85000. It was funded by the Laboratory Directed Research and Development Program of Sandia National Laboratories.

References

1. Masuda, S.; Itagaki, T.; Kosakada, M. *IEEE Trans. on Industry Applications* **1988**, *24*, 740-744.
2. Lee, H.; Williams, S.; Wahl, K.; Valentine, N. *Anal. Chem.* **2003**, *75*, 2746-2752.
3. Mela, P.; van den Berg, A.; Fintschenko, Y.; Cummings, E. B.; Simmons, B. A.; Kirby, B. J. *Electrophoresis* **2005**, *26*(9), 1792-1799.
4. Chou, C.; Tegenfeldt, J.; Bakajin, O.; Chan, S.; Cox, E.; Darnton, N.; Duke, T.; Austin, R. *Biophys. J.* **2002**, *83*, 2170-2179.
5. Zhou, G.; Imamura, M.; Suehiro, J.; Hara, M. *Proc. IEEE: 37th Ann. Meet. IEEE Indust. Appl. Soc.*, Pittsburgh, PA, **2002**; 1404-1411.
6. Suehiro, J.; Zhou, G.; Imamura, M.; Hara, M. *Proc. IEEE: Ann. Meet. IEEE Indust. Appl. Soc.*, Pittsburgh, PA **2003**; 1514-1521.
7. Cummings, E.; Singh, A. *Anal. Chem.* **2003**, *75*, 4724-4731.
8. Lapizco-Encinas, B. H.; Simmons, B. A.; Cummings, E. B.; Fintschenko, Y. *Anal. Chem.* **2004**, *76*, 1571-1579.
9. Lapizco-Encinas, B. H.; Simmons, B. A.; Cummings, E. B.; Fintschenko, Y. *Electrophoresis* **2004**, *25*, 1695-1704.

10. Mills, C. A.; Martinez, E.; Bessueille, F.; Villanueva, G.; Bausells, J.; Samitier, J.; Errachid, A. *Microelectronic Engineering* **2005**, *78-79*, 695-700.
11. Wang, Y. -X.; Cooper, J. W.; Lee, C. S.; DeVoe, D. L. *Lab on a Chip* **2004**, *4*(4), 363-367.
12. R.F. Probstein, *Physico Hydrodynamics*, John Wiley & Sons, 1995.
13. Cummings, E. B.; Griffiths, S. K.; Nilson, R. H.; Paul, P. H. *Anal. Chem.* **2000**, *72*, 2526-2532.
14. Ryuji, Y.; Shoji, T.; Hiroyuki, A. *Analyst* **2004**, *129*(9), 850-854.
15. Craigie, C. J. D.; Sheehan, T.; Johnson, V. N.; Burkett, S. L.; Moll, A. J.; Knowlton, W. B. *J. Vac. Sci. Tech., B: Microelectronics and Nanometer Struc.* **2002**, *20*(6), 2229-2232.

Chapter 10

Design and Synthesis of Dendritic Tethers for the Immobilization of Antibodies for the Detection of Class A Bioterror Pathogens

Charles W. Spangler, Brenda D. Spangler, E. Scott Tarter, and Zhiyong Suo

SensoPath Technologies, Inc., 2100 Fairway Drive, Suite 104, Bozeman, MT 59715

The possibility of bioterror attacks on civilian populations, particularly in densely-populated urban areas, has not abated since the exposure of postal workers to anthrax spores and, possibly, weapons grade anthrax aerosols. Even though there have been a number of potential detection systems suggested for the rapid identification of bioterror pathogens, there are still no easily-used detection systems that can be employed at the point of attack by first responders for the rapid identification of all Class A bioterror pathogens (anthrax, plague, smallpox, botulinum toxin and tularemia) as defined by NIH-NIAID. In this chapter, we will report how antibodies to specific antigens might be immobilized on a variety of biosensor surfaces with unique tether systems that offer a remarkable degree of design flexibility, multivalent surface attachment for increased stability, and variable terminal functionality for the immobilization of antibodies directed against protein toxins expressed by Class A bioterror pathogens.

Introduction

Over the years following the anthrax attacks in 2001–2, a number of simple, field-deployable, single-use, commercial devices for the detection of anthrax, ricin and botulinum toxin have appeared in the marketplace. However, there are no simple detection systems available for the detection of plague, smallpox and tularemia organisms, or instrument systems that are multiplexed for multiple pathogen detection in the field, and can be operated by relatively untrained personnel. This is equally true for secondary response laboratories (emergency rooms, community hospitals, etc.), where a positive identification needs to be made as quickly as possible (< 1 h). Since the exposure symptoms for many bioterrorism agents mimic common infections such as flu, it is extremely important to identify multiple pathogenic agents as quickly as possible so that therapeutic prophylaxis can be started. This is crucial if large numbers of people have been exposed, and quarantine needs to be imposed before they disperse over a wide geographic region. For these reasons, we have been developing a potentially universal methodology revolving around antibody immobilization on biosensor surfaces applicable in many different instrument platforms. This technology focuses on the design and synthesis of immobilization tethers for a variety of surfaces that can incorporate many different bio-conjugation moieties, and be designed with variable flexibility, and incorporate both hydrophilic and hydrophobic segments. In this chapter, we will focus on tethers that immobilize antibodies on gold biosensor surfaces (chips) that can be utilized in commercial Surface Plasmon Resonance (SPR) and Quartz Crystal Microbalance (QCM) instruments.

Design and Synthesis of Dithiol Tethers

Effective detection systems of any type require a robust and flexible bioactive capture agent that can respond to very small concentrations of analytes, whether obtained as crude powder samples at the scene of a purported bioterror attack, or from nasal swabs from possibly infected victims. They should also be both specific and selective, and eliminate (or at least greatly reduce) both false negatives and positives. Antibodies that recognize specific antigens are ideal for this purpose, and can be coupled to a biosensor that recognizes a change in analyte binding. However, one must recognize that antibody immobilization cannot be random, and the nature of the binding site cannot be compromised by the method of attachment to the surface, with the active site preferentially presented to the surrounding solution in an unencumbered fashion. With these criteria in mind, SensoPath Technologies has developed a novel and effective design approach to antibody immobilization that employs monodisperse dendritic heterobifunctional tether capture agents that can be derivatized for attachment as self-assembled monolayers (SAMs) on a variety of biosensor surfaces. For attachment to a gold SPR or QCM surface, we have incorporated a dithiol-terminated dendron at one terminus, and a functionality that can react and immobilize with a monoclonal, polyclonal or recombinant antibody on a gold sensor surface. In Figure 1 we illustrate a typical capture agent that identifies the various "building block" components that can be varied in this construct.

In a typical construct in our laboratories, n = 6 for the hydrophobic chains linking the thiol groups to the gold surface and m can equal 2–10 methylene

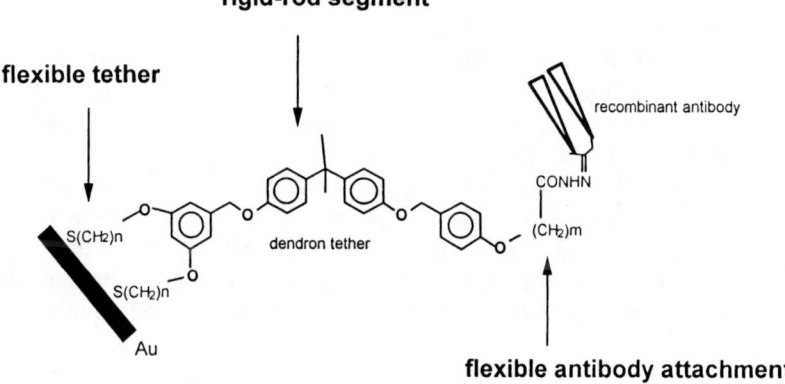

Figure 1. Dendritic dithiol immobilization agents with flexible design controls.

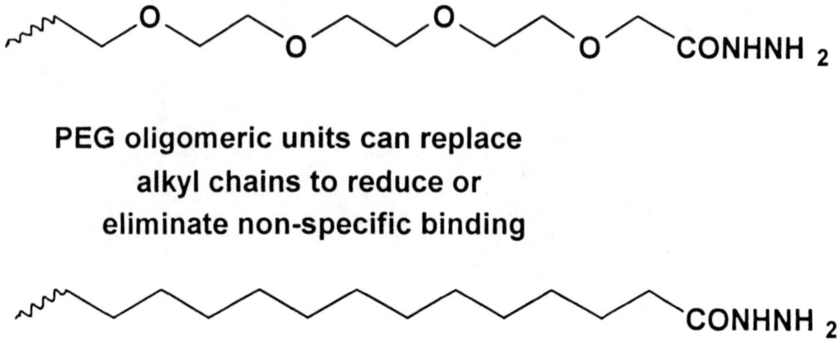

PEG oligomeric units can replace alkyl chains to reduce or eliminate non-specific binding

Figure 2. Choice of hydrophobic or hydrophilic chains in antibody attachment segment.

units, depending on the desired flexibility. Increased flexibility can also be achieved by removing the benzyloxy group linking the bis-phenol A to the antibody attachment segment. Finally, the hydrophobic $-(CH_2)_m-$ portion of this segment can be replaced by a hydrophilic, oligomeric poly(ethylene glycol) (PEG) chain to reduce or eliminate non-specific binding, as shown in Figure 2.

What makes dendritic architecture extremely attractive for the immobilization of bioactive molecules on a sensor surface is its inherent multifunctionality, the capability of building in multiple attachment (e.g., SH) groups, and the capability of attaching a variety of active substrates in a predetermined geometric pattern. For example, the dithiol immobilization construct shown in Figure 1 is essentially built from a G-0 dendron. If this is replaced with a G-1 dendron using well-established Frechét methodology (1, 2), four SH attachment points can be envisioned; however, in this chapter, we will confine our discussion to the dithiol tethers, as they exhibit extraordinary SAM stability compared to the more common monothiol SAMs that have recently been shown to "blink" by a detachment-reattachment mechanism leading to deterioration of the SAM surface (3).

As shown in Scheme 1, alkylation of 3,5-dihydroxybenzyl alcohol with an α,ω-dibromide determines the length of the spacer groups that will eventually terminate in the SH attachment groups, and can be carried out in good yield.

In Scheme 2, we introduce a rigid rod component to provide chain stiffening in the tether mid-section to prevent the construct from folding back on

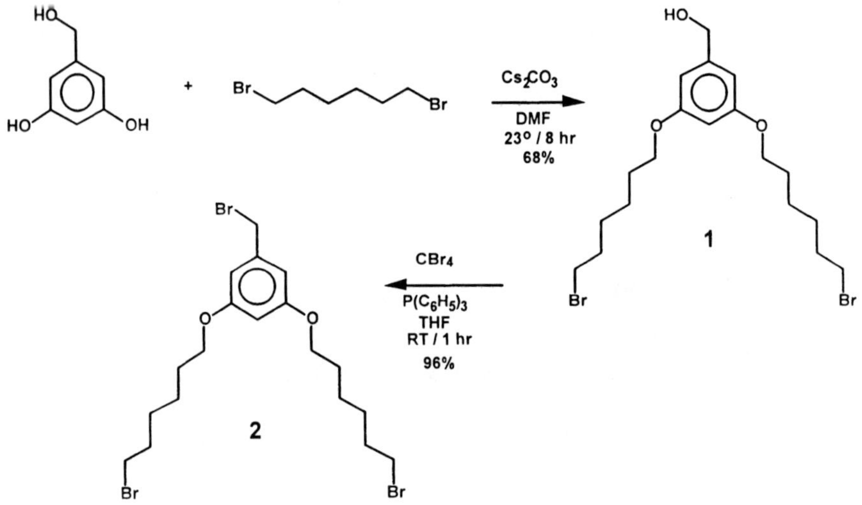

Scheme 1. Synthesis of dendritic tether precursor.

itself. Jepperson and coworkers (*4*) have recently described the impact and significance of the length and configuration of the tether groups on ligand-receptor bond formation in bio-recognition systems. Their detailed analysis, supported by Monte Carlo simulations, diffusion reaction theory, and surface force measurements, of the implications of using long, totally flexible single chain tethers led to the startling conclusion that the tether groups do not exist in the highly extended configurations necessary for binding and that they do not exist as they are usually depicted in model cartoons. While this interpretation dictates that a tether can "stretch" to attach to a receptor, it also points to the possibility that the more rigid dendritic scaffolds we have designed and synthesized may provide a faster, and more effective, binding scenario by eliminating the conformation reordering that is a necessary initial step with more flexible tethers (single chain SAMs).

Finally in Scheme 3, we illustrate how the intermediate **6** can be converted into the final immobilization reagent. First, the terminal bromo groups on the

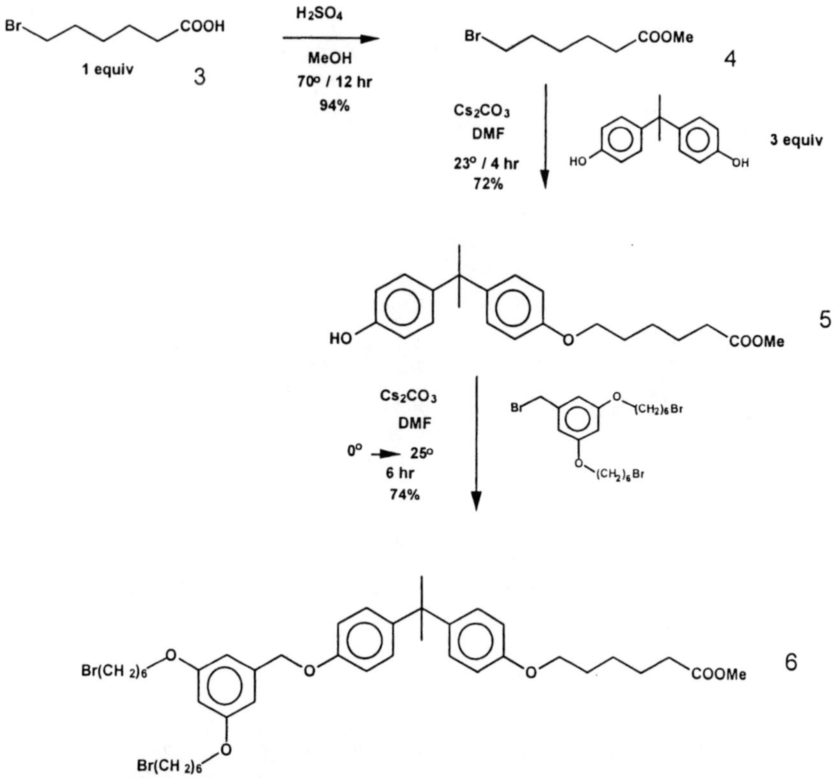

Scheme 2. Synthesis of a functionalized precursor tether incorporating a rigid rod segment.

tether arms are converted into thiol groups for attachment to the gold surface. During this process, the ester group at the opposite terminus is cleaved to the free acid, and, then, subsequently converted back to the ester group, followed by conversion to the desired hydrazide **8** (for coupling to an aldehyde group generated in the antibody by established procedures). The hydrazide will capture polyclonal and some monoclonal antibodies by reacting with aldehyde moieties derived from their attached sugars. Recently, SensoPath Technologies has also developed and described methodology for glycosylating monoclonal and recombinant antibodies and other unglycosylated proteins (5) so that our immobilization approach can make use of any available antibody or protein type (6, 7).

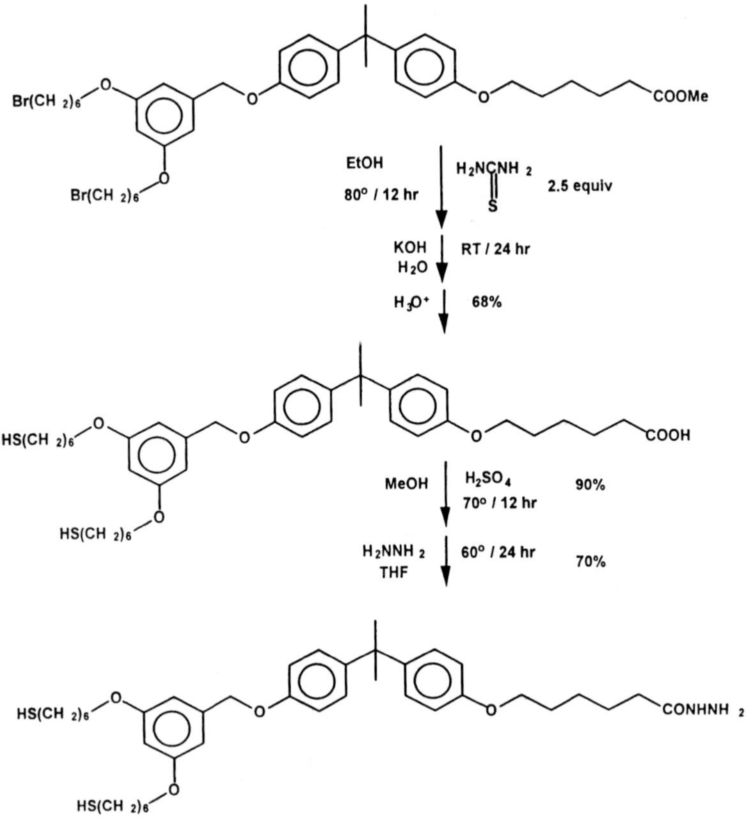

Scheme 3. *Final assembly of the immobilization reagent incorporating dithiol tether groups and antibody immobilization functionality.*

Other functionalities can be incorporated on the detector end of the construct. As shown in Figure 3, COOH, COOR and CONHNH$_2$ are readily synthesized using our methodology, but other groups, such as esters derived from N-hydroxysuccinimide, as well as biotin, can be incorporated using standard protocols.

Rationale and Synthesis of Dithiol Tethers for Enhanced SAM Stability and Reduced Nonspecific Binding

One potential problem for biosensor SAMs when the detection of small concentrations of a bioterror pathogen antigen is required is that of nonspecific absorption. The most commonly-used SAM surfaces in Surface Plasmon Resonance biosensors (SPRs) over the past several years have been commercially-available, carboxymethyl dextran-coated surfaces (CM5 chips, BIAcore), or synthetically-derived, small molecules linked to a monothiol tether group. The dextran chips, though widely used, have the disadvantage associated with nonspecific binding and antibody active site availability. Small molecule-tethered SAMs require some dexterity in synthetic organic chemistry not available in many laboratories, and custom designed SAMs have not been available prior to SensoPath's involvement with this field. Lahiri, et al., have described (8) a unique ensemble poetically described (Knoll) as "flowers in the meadow" (9) that is basically a mixed SAM that is assembled from a tri(ethylene glycol)-terminated alkylthiol ("meadow"), and a longer carboxylic acid terminated hexa(ethylene glycol) alkyl thiol ("flower"). The percentage of either component can be varied from 0.0 to 1.0, however, in most applications involving the detection of large antigens, the "flower" component usually is less than 10% of the mixed SAM. Lahiri, et al. (8) immobilized ten proteins on the mixed SAM by first converting the "flower" COOH group to an NHS (N-hydroxysuccinimidyl ester) leaving group which was subsequently coupled to amine functionalities in the protein. These SAMs proved to be highly resistant to nonspecific absorption over a wide range of molecular weights and isoelectric points. In Figure 3, shown below, we illustrate a typical "flowers-in-the-meadow" SAM as described by Lahiri, et al. (8). It should be emphasized the "flower" component is randomly distributed in the SAM, but gives good surface area coverage for detection of antigens when coupled to antibodies since they are much larger than the "meadow" component.

One problem with the SAMs described above (3) has been alluded to previously in that monothiol tethers can dissociate and reattach to a gold surface. In an SPR flow mode, this can result in gradual loss of SAM component molecules, resulting in "patchy" surfaces and the gradual decay in biosensor performance. We have designed new tether molecules using the "flowers-in-the meadow" concept that result in SAM surfaces with much greater stability, yet still retain the desired low nonspecific binding property of the monothiol counterparts. To a large extent, the design and synthesis of the dithiol tethers follow the syntheses described previously in Schemes 1-3, with the elimination of the rigid rod component and the incorporation of PEG units according to the Lahiri model (*8*).

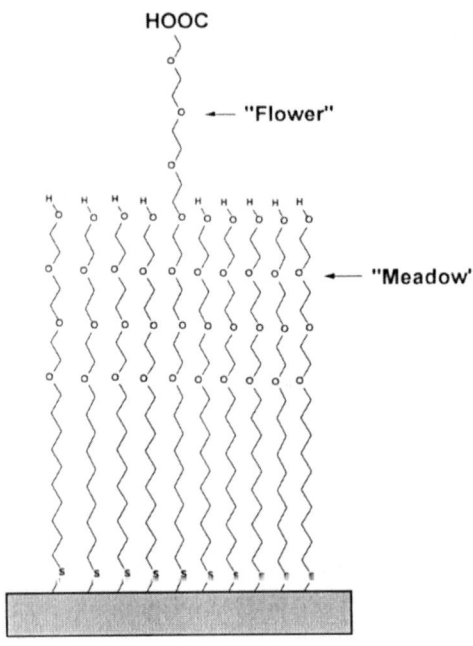

Figure 3. Typical "Flower-in-the-meadow" SAM (after Knoll (9)).

The dithiol counterpart to the monothiol tether illustrated in Figure 3 is illustrated in Figure 4. The synthesis of both "flower" and "meadow" components is illustrated in Scheme 4. The dithiol tethers identified above as "flower" (COOH terminated) and "meadow" (OH terminated) components of the SAMs were synthesized from a common intermediate previously identified as compound **2**. The benzylic bromide in **2** is much more reactive than the two primary bromides on the tether arms, and can be reacted with PEG oligomers to yield an OH terminated PEGylated arm, compounds **10a–d** in approximately 60% yields following chromatographic purification. These, in turn, are reacted with ethyl bromoacetate to give compounds **11a–d** in ca. 60% yields. The thiol substituents are then formed as previously described in Scheme 3, and following basic workup, the functionalized carboxy-terminated PEGylated tethers **12a–d** are obtained in 30% overall yields. It should be noted that the yields for different PEG oligomer lengths are relatively independent of the length of the PEG oligomer. For the current work, the OH terminated meadow was synthesized using triethylene glycol (n = 3) and the COOH terminated tethers contained tetraethylene glycol (n = 4). More recently, we have found that COOH tethers having n = 6 are very effective in mixed SAMs, and the synthesis has now been standardized using that oligomer length. The OH terminated dithiols **13a–d** are formed from the dibromides **10a–d** as previously described in Scheme 3.

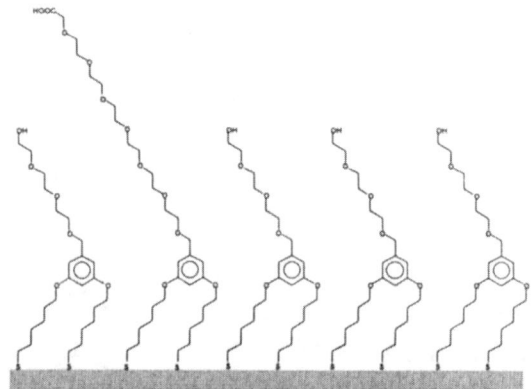

Figure 4. Dithiol "Flowers-in-the-meadow" mixed SAM.

In collaboration with scientists at Reichert, Inc. (*10*), we have recently demonstrated the efficacy of our dithiol "flower-in-the-meadow" SAM design using a surface plasmon resonance biosensor to monitor warfarin binding to immobilized human serum album. The results demonstrate the feasibility of using this surface to reproducibly monitor the binding of a low molecular weight analyte such as warfarin. The surface immobilized a suitable amount of protein (Figure 5), and provided extremely low (undetectable), non-specific binding when challenged with a 1,000-fold excess of non-specific protein over specific antibody (Figure 6).

Non-specific binding characteristics, shown in Figure 6, were determined by injecting a very high concentration (133 nM, 20 µg) Sheep immunoglobulin (IgG), a non-specific protein, over bovine serum albumin (BSA) immobilized on the mixed dithiol, self-assembled monolayer, followed by injection of (0.13 nM, 20 ng) specific anti-BSA immunoglobulin (*10*). The lack of response to the Sheep IgG injection demonstrates the extremely low non-specific binding properties of the dithiol surface. We have extended them by binding anti-anthrax antibody to immobilized anthrax toxin subunit, as well as detection of the toxin subunit by a recombinant anti-anthrax antibody.

We have shown in situ (directly on the gold-coated SPR slide) reaction between a carboxy-methyl-terminated SAM and hydrazine, monitored in real-time by SPR, as reagents were injected into a flow cell. The reaction generated a fresh hydrazide terminus on the SAM that was used to couple an aldehyde-

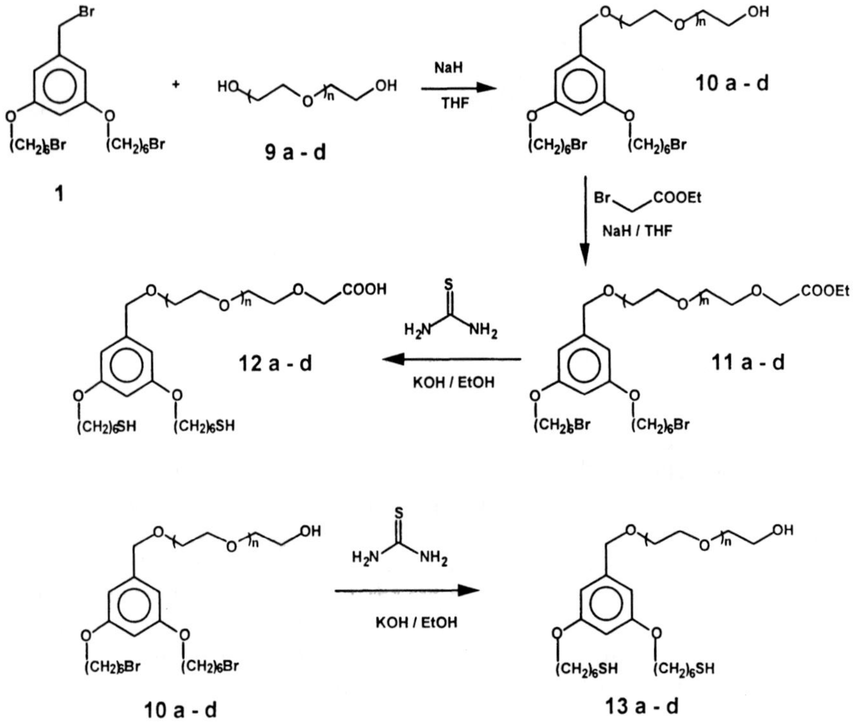

a, n = 2; b, n = 3; c, n = 4; d, n = 5
(13 a = "meadow" ; 12 d = "flower")

Scheme 4. Synthesis of dithiol "flower" and "meadow".

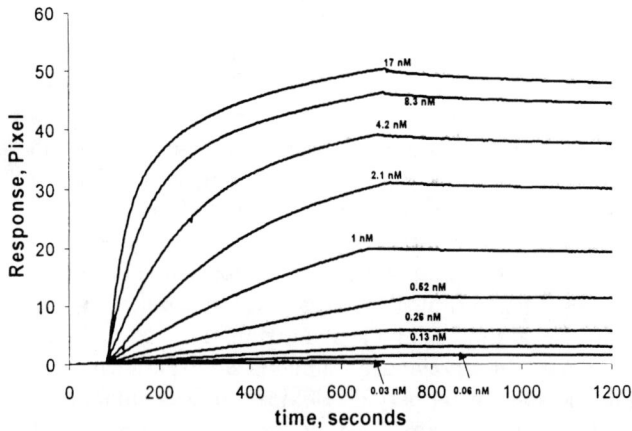

Figure 5. Anti-BSA immunoglobulin (antibody) (0.26–16.7 nM) binding to surface-immobilized BSA.

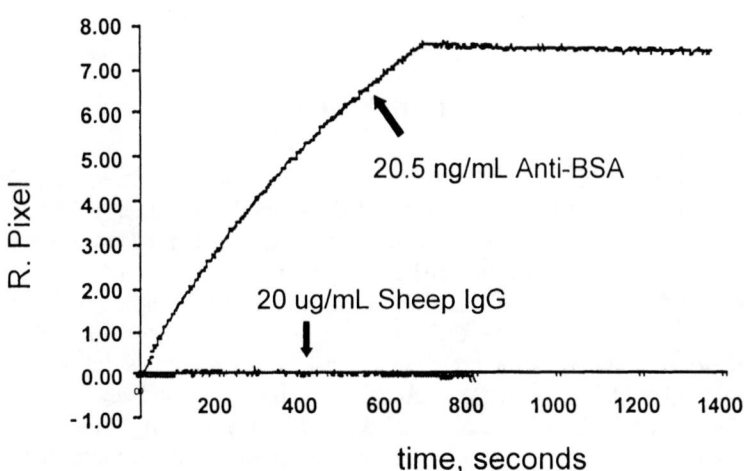

Figure 6. Surface plasmon resonance comparison between specific and non-specific binding to BSA immobilized by the dithiol mixed SAM on gold.

functionalized monoclonal antibody to the hydrazide-terminated tether. The antibody, anti-anthrax protective antigen (PA), recognizes the PA subunit of anthrax toxin. Anthrax PA is required for host cell recognition and subsequent intoxication by associated toxic anthrax enzymes. It is therefore a crucial and unique protein for identification of anthrax by any biosensor device. Antibody was prepared for coupling by periodate oxidation of carbohydrate residues linked to the antibody protein. The resulting aldehyde moiety was coupled covalently to the hydrazide-terminated tether, immobilizing it on the gold surface of the SPR slide in a precise orientation determined by the position of the carbohydrate residues linked to specific asparagines residues on the antibody molecule (*11*). Anthrax PA protein was then introduced. PA was captured by tethered antibody, with very little PA dissociating. The system demonstrates the feasibility of using SPR as a rapid biosensor for detection of pathogenic agents.

We have also demonstrated the feasibility of monitoring antibody titer in serum samples in real-time using the stable dithiol mixed SAM surface coupled to relevant antigen. In a previously unpublished experiment, PA, prepared by glycosylation and subsequent periodate oxidation, was immobilized on a mixed dithiol SAM bearing hydrazide-terminated "flowers". The antigen-functionalized surface was repetitively probed with sequential peak fractions of chromatographically-purified anti-PA recombinant antibody. Bound antibody was dissociated by high-salt buffer between each sample introduction. Six fractions were tested, and the relative concentration of each could be determined. The SPR system can also be used in vaccine development to quickly monitor serum samples during development of the immune response.

Conclusions

In this study, we have described a unique, new tethering system that can immobilize a wide variety of capture agents to gold surfaces in SPR and QCM biosensors. Coupling to other surfaces (e.g., glass) can be envisioned by a simple replacement of the two SH groups (e.g., $-Si(OEt)_3$ for glass or quartz surfaces). These new immobilization agents can also be coupled to a variety of nanoparticles, including gold. Over the next few years, SensoPath Technologies will conduct detailed structure-property relationship studies to optimize the various tether lengths, PEG oligomer incorporation, and the identity and positioning of rigid rod segments. We will also identify other additional coupling agents to extend the variety of bioactive capture agents that can be immobilized on biosensor surfaces. Such studies have been previously carried out for monothiol SAMs by the Whitesides group, and others (*12-15*), and can provide valuable insights into optimization of this new design paradigm. Whatever the results of these studies, it is certain that even better immobilization agents can be synthesized and tested over the next few years.

Acknowledgments

The authors acknowledge partial support of the research results reported in this chapter by a NIH-NIAID Phase I SBIR award.

References

1. *Dendrimers and Other Dendritic Polymers*; Fréchet, J. M. J.; Tomalia, D. A., Eds.; John Wiley & Sons, Ltd.: New York, NY, 2001.
2. *Dendrimers and Dendrons*; Newkome, G. R.; Moorefield, C. N., Eds.; Wiley-VCH: Weinheim, Germany, 2001.
3. Ramachandran, G. K.; Hopson, T. J.; Rawlett, A. M.; Nagahara, L. A.; Primak, A.; Lindsay, S. M. *Science* **2003**, *300*, 1413-1416.
4. Jeppersen, C.; Wong, J. Y.; Kuhl, J. L.; Israelachvili, J. N.; Mullagh, N.; Zaplipsky, S.; Marques, C. N. *Science* **2001**, *293*, 265-268.
5. Hyman, D. M. S. *Thesis*, Montana State University, Bozeman, MT, 2003.
6. Spangler, B.; Hyman, D.; Tarter, E. S.; Spangler, C. *Function-based and other novel approaches to sensors for Homeland Defense, Symposium, Amer. Chem. Soc. Nat. Meeting*, New Orleans, LA March, 2003.
7. Hyman, D.; Spangler, B. *Future directions for bio-defense, Symposium, Amer. Soc. Microbiology*, Baltimore, March, 2003.
8. Lahiri, J.; Isaacs, L.; Tien, J.; Whitesides, G. W. *Anal. Chem.* **1999**, *71*, 777-790.
9. Knoll, W., personal communication.
10. Reichert, Inc., Depew, New York.
11. Spangler, B. D.; Tyler, B. J. *Anal. Chim. Acta* **1999**, *399*, 51-62.
12. Bain, C. D.; Evall, J.; Whitesides, G. M. *J. Am. Chem. Soc.* **1989**, *111*, 7155-7164.
13. Bain, C. D.; Whitesides, G. M. *J. Am. Chem. Soc.* **1989**, *111*, 7164-7175.
14. Pale-Grosdemange, C.; Simon, E. S.; Prime, K. L.; Whitesides, G. M. *J. Am. Chem. Soc.* **1991**, *113*, 12-20.
15. Schlenoff, J. B.; Li, M.; Ly, H. *J. Am. Chem. Soc.* **1995**, *117*, 12528-12536.

Decontamination and Protection

Chapter 11

Amphiphilic Polymers with Potent Antibacterial Activity

M. Firat Ilker[1,2], Gregory N. Tew[1], and E. Bryan Coughlin[1]

[1]Department of Polymer Science and Engineering, University of Massachusetts, Amherst, MA 01003
[2]Current address: Department of Chemistry, University of Wisconsin, Madison, WI 53706

> A general strategy is summarized for the assembly of polar and nonpolar domains into a modular monomer structure. Living ring-opening metathesis polymerization (ROMP) of these monomers provides access to a large range of molecular weights with narrow molecular weight distributions. The character and size of each domain can be tuned independently and locked into the repeating unit of the amphiphilic polymers. Lipid membrane disruption activities were investigated for amphiphilic polynorbornene derivatives against liposomes. Water-soluble, amphiphilic, cationic polynorbornene derivatives, which exhibited the highest level of activities against liposome membranes, were then probed for their antibacterial activities in growth inhibition assays and hemolytic activities against human red blood cells in order to determine the selectivity of the polymers for bacterial over mammalian cells. By tuning the overall hydrophobicity of the polymer, highly selective, non-hemolytic antibacterial activities were obtained. These simple polymers represent a new approach to the development of nontoxic, broad-spectrum antimicrobials and have significant potential for applications in bio-terrorism defense.

© 2008 American Chemical Society

Introduction

Well-defined amphiphilic macromolecules find important applications in biology and medical sciences. Examples include the use of polymeric materials in drug delivery (*1–5*), gene delivery (*6–9*), tissue engineering (*10–12*), and antibiotic agent applications (*13–19*). Continuing research efforts are focusing on the use of polymeric therapeutics as alternative antibiotic agents in the fight against bacterial diseases. Antibacterial activity of cationic polymers has been known for several decades (*14*). Various polymeric structures carrying cationic moieties have found considerable interest in non-medical use, such as food preservatives, pesticides, and disinfectants (*13*). Very recently, antibacterial activity of relatively simple cationic polymers has started to be considered within the scope of the studies involving naturally occurring host-defense peptides, and their synthetic mimics (*20–22*). Although more complex in their structure, antimicrobial peptides commonly contain cationic and hydrophobic domains (*23*). Successful research efforts that target synthetic mimics of host-defense peptides have typically followed a top-down approach, through modifications of naturally occurring peptide structures, in an effort to establish an understanding of structure-property relationships (*24*). In the development stage, synthetic mimics of host-defense peptides require elaborate and extensive techniques (*25–28*). Relatively simple synthetic cationic polymers offer an inexpensive alternative; however, they suffer from their high cytotoxicity if considered for therapeutic applications (*13*). Encouraged by the synthetic abilities for the controlled preparation of amphiphilic polymers, and inspired both by antimicrobial peptide research and synthetic biocidal polymers, we seek the determination of macromolecular properties that allows for antibacterial activity, while suppressing cytotoxicity. This chapter is a compilation of our efforts, including the design and synthesis of novel amphiphilic polymers (*29*), probing their structure-biological activity relationship patterns, and, finally, optimization of their amphiphilic character for selective antibacterial activity (*30*).

The initial focus of the current study is the preparation and homopolymerization of a novel class of monomers with amphiphilic character, where the amphiphilicity of the resulting polymer is tuned at the repeating unit level, giving rise to a polymer backbone structure with regularly spaced hydrophilic and hydrophobic groups. The molecular weight of the amphiphilic polymer is independently controlled through the choice of polymerization procedure. Ring-opening metathesis polymerization (ROMP) of amphiphilic modular monomers will be described as a synthetic tool for the facile probing of the effect of basic macromolecular variables on the interactions of polymers with living cells, prokaryotes and eukaryotes (*29*).

The starting point for monomer design is based on widely used norbornene derivatives. Norbornene derivatives having 2-mono or 2,3-difunctionalization are known to be excellent monomers for ROMP. They have been used in the preparation of a wide range of polymeric structures (*31*). Using various norbornene derivatives, polymers bearing a variety of side groups have been prepared via ROMP. Functionalized norbornene derivatives are readily prepared via Diels-Alder cycloaddition of a diene, most generally furan or cyclopentadiene, to a dienophile possessing a desired functional group (*31*). This procedure affords an endo or exo 2- or 2,3-functionalized norbornene derivative. Endo isomers are known to be poor monomers for ROMP, presumably because of the increased steric crowd around the polymerization-active carbon-carbon double bond.

The task of preparing a monomer structure with dual functionality, in this case a hydrophilic and a hydrophobic group, lead us to investigate the preparation and polymerization of modular norbornene derivatives with additional functionality at the 7 position of the ring (Figure 1). Using this general strategy, two complementary functionalities can be introduced into the monomer structure and the properties of the resulting amphiphilic polymer can thus be fine-tuned.

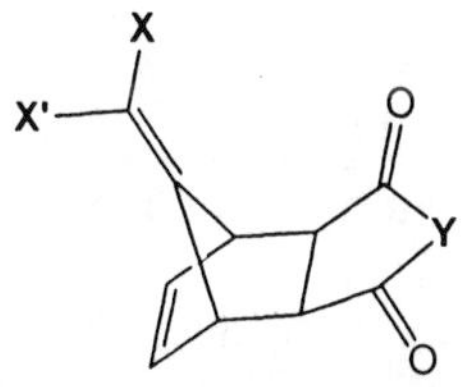

X, X' = hydrophobic
Y = hydrophilic

Figure 1. General structures of amphiphilic modular norbornene derivatives.

Monomer Synthesis

Fulvene derivatives were used as functionalized dienes for the Diels-Alder cycloaddition reaction with an appropriate dienophile to obtain the modular norbornene structures (Figure 2). Three different fulvene derivatives, 6,6'-dimethyl fulvene, 6-isopropyl fulvene, and 6,6'-di-*n*-propyl fulvene, were prepared through a simple, synthetic methodology involving pyrrolidine-catalyzed condensation of cyclopentadiene with the appropriate aldehyde or ketone, resulting in high yields (*32*). The hydrophobic character of the monomer and the resulting polymer can be tuned by the choice of fulvene derivative. The modular approach to the monomer preparation allows for a variety of different alkyl groups to be readily incorporated. This allows for facile increase, or decrease, of the hydrophobic character of the monomer and, thus the resultant polymer.

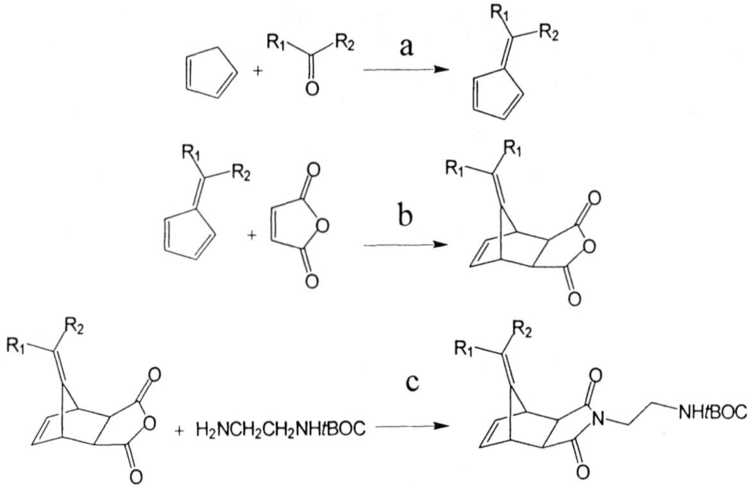

Figure 2. Representative preparation of modular norbornene derivatives. (a) pyrrolidine, CH_3OH (b) EtOAc, 80 °C (c) $CoAc_2$, Ac_2O, DMAc, 80 °C, 4 h.

Maleic anhydride was used as the dienophile, allowing for further functionalization following the assembly of the norbornene skeleton. Diels-Alder cycloadditions of the above-mentioned fulvene derivatives with maleic anhydride at elevated temperatures, between 80 °C and 120 °C, and moderate concentrations, 0.2 to 0.5 M, afforded quantitative yields of the corresponding norbornene derivatives **5**, **7**, and anhydride precursor of **9** (see Table I) (*29*). At total adduct concentrations above 1.5 M or temperatures above 130 °C, a solid oligomeric side product, presumably a copolymer of the reactants, was obtained. Two isomers, endo or exo, can be obtained from cycloaddition reactions, depending on the nature of adducts or the reaction temperature. These isomers exhibit different polymerization kinetics, where, in most cases, endo adducts polymerize very slowly, and result in low conversions.

To achieve a high-level of control over polymerizations, and resulting polymer microstructures, the preparation of pure exo isomers of the monomers were targeted. Exo-endo mixtures that were obtained as the cycloaddition adducts were not always separable by selective recrystallizations. Compounds **5**, **6**, **8**, and **9** were separated from their endo isomers through selective recrystallization to give white crystalline solids. Cobalt-catalyzed transformation of the anhydride into a substituted imide linkage resulted in the protected amine functionalized monomer structure in excellent yield. For monomers **6**, **8** and **9**, pure exo isomer was isolated by successive recrystallizations from cold ether; overall yields were 40 to 56%.

Table I. Examples of modular norbornene derivatives and amphiphilic polymers resulting from corresponding polymerizations[a] and deprotections

Monomer	Deprotected Polymer	Theo. M_n[b] (g/mol)	Obs. M_n[c] (g/mol)	PDI[c]
5	dep-poly5	2,000	2,900[d]	1.15[d]
		5,100	7,000[d]	1.14[d]
		10,200	10,000[e]	1.17[e]
6	dep-poly6	2,400	1,950	1.13
		5,900	7,000	1.08
		19,800	17,900	1.11
		29,900	24,100	1.13
7 (exo-endo)	dep-poly7	10,900	9,500	1.63
		21,900	19,500	1.49
8	dep-poly8	1,900	1,800	1.20
		8,800	8,600	1.10
		31,100	27,000	1.13
		63,300	57,200	1.70
9	dep-poly9	4,900	5,300	1.09
		14,600	14,500	1.24
		32,300	32,200	1.13
		60,500	57,000	1.19

[a]Polymers were prepared using catalyst 4 (see Figure 3). [b]Theoretical molecular weights calculated based on catalyst to monomer ratio, assuming full conversion. [c]Determined by THF GPC relative to polystyrene standards prior to deprotection of polymer. [d]Determined by water GPC relative to poly(ethylene oxide) standards. [e]Determined by DMF GPC relative to polystyrene standards prior to the hydrolysis of polymer.

Homopolymerization Studies

The initial target of the current study was to prepare amphiphilic polymers with well-defined architectures. Because the amphiphilic character was already dictated in the monomer unit, the target in the polymerization study of modular norbornene derivatives was to achieve controlled polymerization and obtain narrow polydispersities. The polymerization of a model monomer, **8**, was tested using four different metathesis catalysts, **1–4** (see Figure 3), in order to screen the polymerizability and the effect of catalyst on the resulting polydispersities. The polymerization of **8** using catalysts **1–3** required elevated temperatures between 40–55 °C; whereas, catalyst **4** allowed for polymerization at room temperature. This result was in accordance with the reported high reactivity of catalyst **4** (*33*). Desired molecular weights ranging between 1,600 g/mol to 75,000 g/mol (M_n) could be obtained by adjusting the catalyst to monomer ratio for all four catalysts. For a targeted number average molecular weight of 8,800 g/mol at complete conversion, the polymerization of **8** using catalysts **1–4** resulted in polydispersity values of 1.23, 1.27, 1.96, and 1.10, respectively. Based on these results, the homopolymerizations, and subsequent random and block copolymerizations, involving monomers **5–9** were studied using catalyst **4** (Table I). Polymer obtained from monomer **5** (poly5) precipitated from the polymerization solution. Despite the early precipitation during polymerization, 88 to 90% yield of poly5 was isolated, with polydispersity values ranging between 1.14 and 1.17 (M_n ranging from 2,900 to 10,000 g/mol). From the polymerization of monomer **6** using catalyst **4**, poly6 was obtained in 85 to 90% yield, with polydispersity values ranging between 1.08 and 1.13.

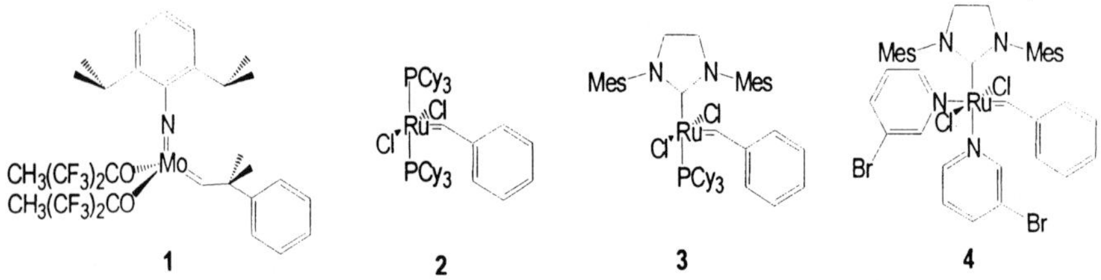

Figure 3. Catalysts 1–4.

For all monomers, the obtained molecular weights were in agreement with the targeted molecular weights, as observed from GPC results. The slight discrepancy between the targeted and observed molecular weights of the polymers in Table I was expected due to the differences in hydrodynamic volume of these polymers versus narrow polydispersity polystyrene and poly(ethylene oxide) GPC standards. ^{1}H NMR end-group analysis was performed to confirm the match between the targeted and observed number average molecular weights for samples with M_n values less than 9000 g/mol. The relative integrations of the resonances from the repeat units versus the multiplet from styrenic end-group at 7.32 ppm were in good agreement with the targeted molecular weight.

The polymerizations of exo-endo mixtures resulted in very low yields, where only a fraction of the exo isomer was polymerized. Despite the presence of endo isomer in the case of monomer 7, the exo-endo mixture was polymerized in good yields into high molecular weight polymers using catalysts 3 and 4; however, the resulting polydispersities were broader when compared to the other monomers (Table I).

Polymer Deprotection to Form Polyelectrolytes

The *t*-BOC protected pendant primary amine groups of poly6, poly8, poly9, poly10, and the anhydride functionalities of poly5 and poly7 provide a non-ionic and hydrophobic character to these polymers that allows for controlled ROMP and subsequent characterization of the polymers in a wide range of organic solvents. To obtain the final amphiphilic nature of the polymers, these groups were deprotected into their ionic forms, resulting in water-soluble polymers. Protected primary amine functionalities of different molecular weight samples of polymers were deprotected, quantitatively, by dissolution in warm (45 °C) trifluoroacetic acid (TFA) to obtain dep-poly6, dep-poly8, dep-poly9, and dep-poly10, as observed by ^{1}H NMR recorded in D_2O solutions. ^{1}H NMR spectra of these polymers also showed that carbon-carbon double bonds on the polymer backbone remained unaffected after treatment with TFA. Anhydride functionalities of poly5, and poly7 were hydrolyzed by dissolution of polymers in NaOH solutions to obtain dep-poly5, and dep-poly7. After these processes, well-defined amphiphilic polymers with a desired anionic or cationic character and hydrophobic character were obtained.

Study of Phospholipid Membrane Disruption Activities

One interesting aspect of amphiphilic macromolecules is their interactions with phospholipid membranes, natural or artificial. Also amphiphilic in their chemical nature, phospholipid building blocks change their supramolecular ordering by incorporating amphiphilic polymers within their membrane assemblies. Depending on structural and compositional factors, various membrane deformations such as pore or tube formation, or complete disruption have been reported (*34–40*). In this respect, biological activities of amphiphilic polymers are often associated with their ability to permeate cell membranes. Because phospholipid-based cell membranes are a principal structural component of living organisms that exclude the interior from the outside environment, their disruption by amphiphilic polymers and oligomers has attracted attention in the biomedical field. Applications, which are based on polymer-induced transport through or disruption of cell membranes, include drug delivery (*1–3*), gene delivery (*6–8*) and antibacterial agents (*20, 27, 41–44*). The antibacterial activity of cationic amphiphilic macromolecules, which will be discussed in the following sections, has been suggested to be through perturbation of bacterial cell membranes (*23, 34, 45, 46*).

Similarly, toxicity against mammalian cells can also be induced by the disruption of cell membranes, often measured as hemolytic activity against red blood cells (*24, 25, 27, 44*). The difference in the lipid compositions of cell membranes from different organisms has been widely suggested to be one of the likely causes for the selective activities of certain membrane disrupting antibacterial agents (*24, 34, 41*). Bacterial cell membranes are known to contain an excess of negative charge on the polar outer surface of their cell membranes. Mammalian cell membranes, on the other hand, possess a neutral zwitterionic outer surface, and contain cholesterol that stiffens the membrane. The outcome from the exposure of phospholipid membranes to amphiphilic macromolecules is dictated by the detailed physiochemical properties of both parties (*24, 47, 48*).

These scientific findings and suggestions, summarized above, were elucidated by a large number of studies that commonly utilize artificial liposomes as model membranes (*28, 34, 40, 44, 47, 49–52*). Liposomes consist of a phospholipid bilayer envelope isolating an inner volume (*53, 54*). They are available through well-established preparative techniques that allow strict control over molecular components of the membrane and the environment. Depending on the preparation details, the average diameters of vesicles typically range between 0.1–5 μm, with a lipid bilayer thickness of several nanometers. With these structural features, liposomes have also been widely studied as microcapsules for drug and gene delivery applications (*55–58*). Liposomes make it possible to monitor the dynamics of membrane perturbations, either by observing deformations using microscopy, if applicable, or by using an appropriate fluorescent dye encapsulated within, or excluded from, the liposome (*59–61*). The leakage of the fluorescent dye can be monitored as an indication of increased permeability or disruption of the membrane (Figure 4). In a typical experiment, liposomes are loaded with a self-quenching concentration of fluorescent dye. Following the addition of membrane-disrupting agent, the disruption of vesicles can be monitored by quantitatively measuring the increasing fluorescence arising from the leakage and dilution of the dye in the larger outer volume.

The disruption of neutral or anionic liposomes, with respect to their total lipid content and surface charge, has been commonly correlated to the selective activities of certain antibacterial agents against bacterial versus mammalian cells

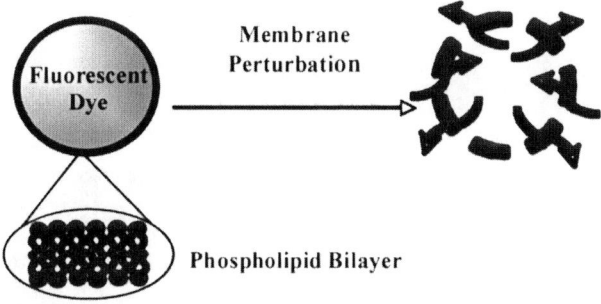

Figure 4. Representative illustrations of liposome and dye leakage experiment.

(28, 40, 44, 47, 49–52). These assays are well-documented in the literature and provide useful insight about the structure-property relationships of membrane-disruptive agents *(20, 27, 44)*. Neutral zwitterionic liposomes, as mimics for mammalian cell membranes, are typically prepared from mixtures of 1-stearoyl-2-oleoyl phosphatidylcholine (SOPC) and cholesterol (CL) as a minor component (Figure 5) *(44)*. Anionic liposomes, on the other hand, are prepared from SOPC and anionic phospholipid 1-stearoyl-2-oleoyl phosphatidylserine (SOPS), as mimics for bacterial cell membranes *(34, 39, 40)*. These are simplified abiogenic models and, therefore, in our study, these tests were used to evaluate the overall membrane disruption activities of polymers. We do not make direct comparisons of these results to activity against biological cells. Selective disruption activity against anionic liposomes or neutral liposomes often depends on very subtle structural details of the amphiphilic macromolecule. The level of structural control over amphiphilic polynorbornene derivatives is used to control lipid membrane disruption activities. The effects of hydrophobicity and molecular weights for these amphiphilic polymers are probed against liposomes of different lipid content *(29)*.

The current study was conducted side by side with the synthetic efforts presented above. Dep-poly**8**, which possesses an intermediate hydrophobicity compared to dep-poly**6** and dep-poly**9**, was the first to be probed for its membrane disruption activities against a series of neutral and anionic liposomes. The effects of lipid content, polymer concentration, and molecular weight on the outcome of membrane disruption activities were elucidated using dep-poly**8**. Then, the effect of polymer hydrophobicity was probed by testing dep-poly**6** and dep-poly**9**, with decreased and increased hydrophobicity, respectively.

When anionic liposomes prepared from 1:9 (molar ratio) mixtures of phosphatidylserine (anionic) and phosphatidylcholine (zwitterionic) were exposed to dep-poly**8**, a 13,500 g/mol (M_n) sample caused 100% lysis at concentrations as low as 5 μg/mL. Approximately 50% lysis was observed at a

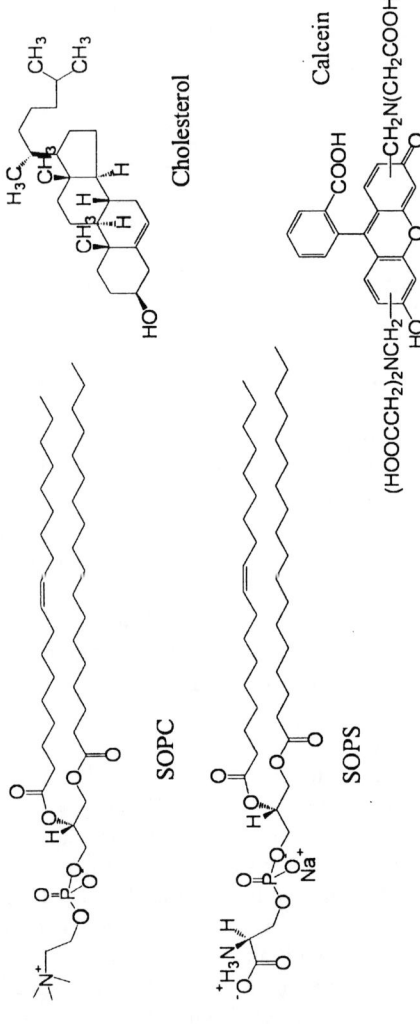

Figure 5. Structures of stearoyl-oleoyl-phosphatidylcholine (SOPC), stearoyl-oleoyl-phosphatidylserine (SOPS), cholesterol, and calcein used in the preparations of liposomes.

concentration of 1.25 µg/mL. Lysis was dose and molecular weight dependent. Figure 6, shows the increase of fluorescence from calcein release in the first 3 minutes after polymer addition, marked as percent lysis, indicating the disruption of vesicles caused by different molecular weights of dep-poly8. When a series of molecular weights of dep-poly8 ranging between monomer and 64,000 g/mol (M_n) were studied, it was observed that the membrane disruption activity was lower for the monomer and oligomers with molecular weights less than 4,500 g/mol. Dep-poly8 of molecular weights above 4,500 g/mol and up to 64,000 g/mol showed very high activities independent of molecular weight in this range. This result suggests that the membrane disruption activity of dep-poly8 increases with molecular weight until it reaches a critical molecular weight necessary to obtain maximum membrane disruption activity. The living nature of ROMP allows for the precise targeting of the desired molecular weight and hence allows for tuning the membrane activity of dep-poly8s.

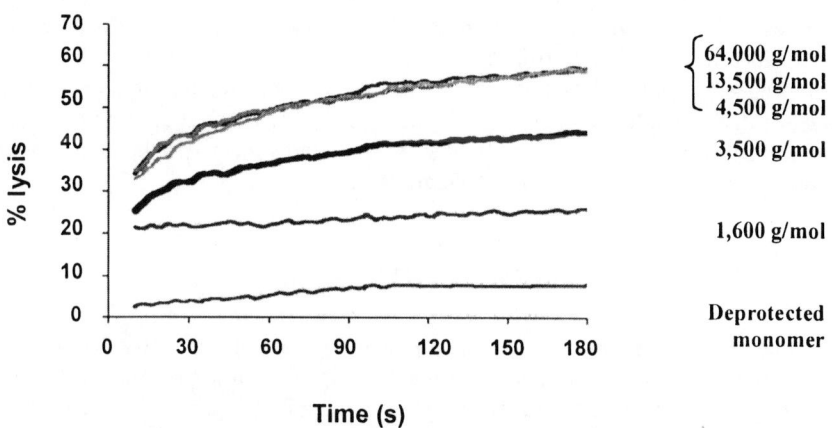

Figure 6. Lysis of anionic liposomes caused by different number average molecular weight (M_n) samples of dep-poly8 at the concentration of 1.25µg/mL.

It should be noted that while probing the molecular weight effect, the concentration of the polymer added into the liposome suspension was calculated in terms of mass/volume. If the corresponding molar concentrations were to be calculated, the dep-poly8 sample with a number average molecular weight of 64,000 g/mol would have 14 times fewer, but longer, chains than the dep-poly8 sample of 4,500 g/mol at the same mass/volume concentration. With this idea in mind, very similar activities obtained from different molecular weights of dep-poly8 at the same mass/volume concentration may suggest a cooperative action of polymer chains being responsible for membrane disruption.

Effect of Membrane Composition of Liposomes

In order to observe the effect of lipid composition of the membranes on the activity of the dep-poly8, neutral, zwitterionic liposomes with a 9:1 SOPC to cholesterol molar ratio, and anionic liposomes with 9:1 and 1:1 SOPC to SOPS molar ratios were prepared. Batches of liposomes with different ionic character were tested within the same experiment, using the same reagents and equipment in order to minimize experimental errors. The results were also confirmed in a second set of experiments.

When a solution of dep-poly8 (M_n = 27,000, PDI = 1.13) in TRIS saline buffer (pH 6.5) was added to each of three different liposome suspensions, the membrane disruption activity was observed to increase with increasing anionic lipid content of liposome from 0 mol % to 50 mol %. 20 μg/mL of dep-poly8 caused 90% lysis in 3 min against anionic liposomes with 1:1 SOPC to SOPS ratio. The percent lysis values at the same experimental conditions decreased to 67% as anionic lipid content decreased to a 9:1 ratio, and 24% for neutral vesicles with no anionic lipid content, but 10% cholesterol content. These results show increasing affinity of dep-poly8 for negatively charged liposome membranes. This trend is consistent with the cationic nature of the polymer causing stronger interactions between phospholipid membrane and the polymer. More than two-fold selectivity against anionic liposomes can be induced by introducing 10% or more anionic lipid content.

Effect of Polymer Hydrophobicity

The previous sections have shown that high disruption activities against phospholipid membranes can be obtained from dep-poly8, depending on polymer molecular weight and membrane composition. The activities of dep-poly6 and dep-poly9 with relatively lower and higher degrees of hydrophobicity were tested against neutral (SOPC: CL = 9:1) and two different anionic (SOPC: SOPS = 9:1 and 1:1) liposomes. Similar molecular weight samples of dep-poly6 (M_n = 24,100, PDI = 1.10), dep-poly8 (M_n = 25,500, PDI = 1.17), and dep-poly9 (M_n = 32,200, PDI = 1.17) were compared within the same experiment (Figure 7). The membrane disruption activities of all polymers were observed to increase with increasing anionic strength of the membrane. However, all three types of membranes were less vulnerable to both dep-poly6 and dep-poly9 when compared to dep-poly8. Increasing or decreasing the hydrophobicity in reference to dep-poly8 resulted in diminished membrane disruption activities.

When the effect of molecular weight was probed for dep-poly6 against anionic liposomes, increased molecular weights were shown to have increased activities, in accordance with the result obtained from dep-poly8. However, for dep-poly6, a plateau of maximum membrane disruption activity was reached over approximately 50,000 g/mol. Finally, when the hydrophobicity was totally removed, in the case of dep-poly10 (M_n = 25,000 g/mol), with an oxygen atom replacing the alkylidene group, the activity against anionic vesicles (SOPC: SOPS = 9:1) was no more than 8% lysis, up to a sufficiently high polymer concentration, 200 µg/mL, within 3 min. These results reveal that a specific hydrophilic/hydrophobic balance is crucial to obtain the highest membrane disruption activities from amphiphilic cationic polynorbornene derivatives.

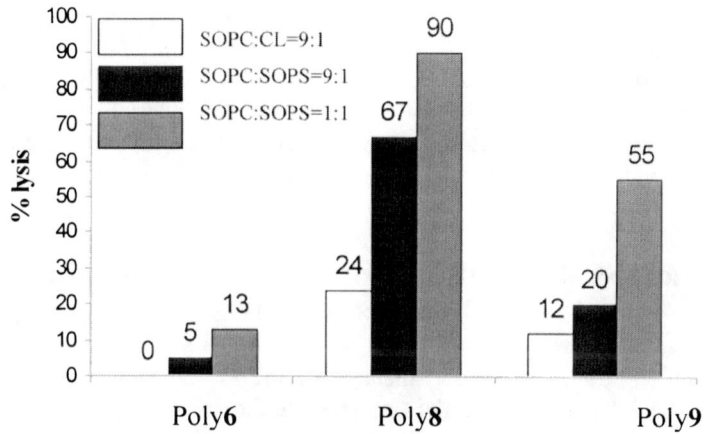

Figure 7. Lysis values 3 min after the addition of 40 µg/mL of dep-poly6 (M_n = 24,100, PDI = 1.10), dep-poly8 (M_n = 25,500, PDI = 1.17), and dep-poly9 (M_n = 32,200, PDI = 1.17) into suspensions of neutral (left, SOPC: CL = 9:1)and anionic liposomes (middle, SOPC: SOPS = 9:1 and right, SOPC: SOPS = 1:1).

Control Experiments

In a control experiment, the anionic dep-poly7 (M_n = 22,000 g/mol) was tested against the liposome suspensions and no lysis was observed at comparable concentrations. It is remarkable that dep-poly7, which has the same hydrophobicity as the cationic dep-poly8, caused no lysis. A change from cationic to anionic character of the polymer resulted in a dramatic decrease of membrane disruption activity. Three commercially available cationic polymers, polyallylamine (M_n = 25,000 g/mol), polyethyleneimine (M_n = 400,000 g/mol), and poly(dimethyldiallyl ammonium dichloride) (M_n = 75,000 g/mol), were also tested as control experiments. These polymer samples provide models for primary amine, secondary and tertiary amine (hyperbranched PEI), and quaternary amine containing polymers, within a large range of high molecular weights. These polymers were observed to be far less active in the lysis of the liposomes when compared to dep-poly8, where percent lysis caused by this set of polymers remained below 10% over 3 minutes at a concentration of 15 μg/mL. These results confirmed once again that cationic amphiphilic polymer structures with a specific hydrophobicity have the highest activity for disruption of phospholipid membranes amongst the polymers studied, thus confirming the overall design principles.

Liposomes provide simplified models for bacterial and mammalian cell membranes, although they underestimate several factors such as cell walls and lipopolysaccharides in bacterial cell membranes. However, our results from membrane disruption activities of amphiphilic polymers built a strong foundation for structure-property relationships of these materials and warrant further exploration of antibacterial activities as well as any other relevant biomedical application of these polymeric materials.

Applications for Well-Defined Amphiphilic Polymers: Antibacterial Activity

Antibacterial activities of macromolecules, including oligomeric compounds, have been studied under two separate thrusts. One group of studies has focused on the structure-property relationships of natural host-defense peptides derived from multicellular organisms (*23, 24, 41, 62*). Despite their structural diversity, most are cationic peptides with a certain degree of hydrophobicity. Extensive studies on the mechanism of action suggest that antimicrobial peptides act by rendering permeable the cell membranes of microorganisms through favorable interactions with negatively charged and hydrophobic components of the membranes followed by aggregation and subsequent disruption (*23, 34, 42, 46*). Host-defense peptides and their synthetic analogs are reported to exhibit varying degrees of activity against wide spectrum of bacteria and mammalian cells (*23*). While host-defense peptides may show selectivity against the membranes of microbes versus the host organism, a number of them are antibacterial and not toxic to human cells, within certain concentration limits, and are thus considered as potential therapeutic agents (*23, 41, 62*). Hemolytic activity against highly susceptible human red blood cells, as representatives of normal mammalian cells, is conventionally used as a measure of cytotoxicity (*24, 34*). Studies aimed at understanding the structure-property relationships of natural peptides have recently evolved into a number of research efforts targeting the preparation of synthetic mimics of antimicrobial peptides. These include stereoisomers of natural peptides (*34, 63*), α-peptides (*47*), β-peptides (*27, 42, 64, 65*), cyclic α-peptides (*26*), peptoids (*28*), and poly-arylamides (*49*), all of which are oligomeric with molecular weight below 3,000 g/mol. Many of these examples target an amphiphilic secondary structure, typically helical, in addition to their cationic nature. Depending on the type of peptide, a facially amphiphilic structure results in the gain, or loss, of selective activity, which reveals that a stable amphiphilic secondary structure is not a precondition for selective antibacterial activity (*34, 47, 65*).

Independent from the antimicrobial peptide research, a second thrust involves studies of synthetic cationic polymers that exhibit varying degrees of antibacterial activities (*13–19*). This class of polymeric compounds is relatively inexpensive and less cumbersome to prepare when compared to peptide mimics. This class of cationic polymers was predominantly targeted for use in the solid state as potent disinfectants, biocidal coatings or filters, due to their toxicity to human cells at relatively low concentrations which is an important distinction from the work on peptide mimics (*13, 14*). Consistent with the targeted applications of these cationic polymers, in most cases, only antibacterial activity was reported without any report of hemolytic activity. In one instance, a soluble pyridinium polymer was reported to have low acute toxicity against the skin of test animals (*66*). An example of antibacterial cationic polymers that have found large industrial use as disinfectants and biocides is poly(hexamethylene biguanide) (PHMB). Different levels of toxicity against various mammalian cells were reported for PHMB and similar biguanide functionalized polymers (*67–71*). To the best of our knowledge, a direct comparison of antibacterial and hemolytic action has not been reported for either of these classes of antimicrobial polymers. Gelman et al. have recently reported the antibacterial activity of low molecular weight, hydrophobically modified, cationic polystyrene derivatives in comparison with a potent derivative of magainin II (*21*). In their initial study, a crossover between the research on antimicrobial peptide mimics and polymer disinfectants, cationic polystyrene derivatives have shown similar antibacterial activities as the magainin derivative, but were highly hemolytic. As a part of very recent efforts in the area, selective activities of facially amphiphilic low molecular weight polyphenyleneethynylenes were reported, with activities and selectivities similar to a magainin derivative (*22*). The successful design of non-hemolytic, antibacterial, and high molecular weight polymers has not been achieved thus far.

Here we present the antibacterial and hemolytic activities of narrow polydispersity homopolymers of modular norbornene derivatives, spanning a large range of molecular weights. The results show that by controlling the hydrophobic/hydrophilic balance of water-soluble amphiphilic polymers, it is possible to obtain high selectivity between antibacterial and hemolytic activities, without a predisposed amphiphilic secondary structure as part of the synthetic design. The overall efficacy toward both Gram-negative and Gram-positive bacteria is strongly dependent on the length of alkyl substituents on the repeat units. The results show that it is possible to design simple polymers that are both potent against bacteria, but non-hemolytic.

Antibacterial and Hemolytic Activities

The hydrophobicity of the repeating unit was observed to have dramatic effect on antibacterial and hemolytic activities of the amphiphilic polymers. The activity of each homopolymers with similar molecular weights (near 10,000 g/mol, M_n) was probed against Gram-negative bacteria (*E. coli*), Gram-positive bacteria (*B. subtilis*), and human red blood cells (Table II). The upper limit of polymer concentration that was required to cause 50% hemolysis is reported as HC_{50}. Antibacterial activity was expressed as minimal inhibitory concentration (MIC), the concentration at which 90% inhibition of growth was observed after 8 h (*30*). Dep-poly10, a cationic polymer with no substantial hydrophobic group, did not show any significant antibacterial or hemolytic activity within the measured concentrations. At the highest concentration measured for hemolytic activity, 1000 μg/mL, dep-poly10 caused 5% hemolysis.

Table II. Antibacterial and hemolytic activities of homopolymers

Polymer	M_n (g/mol)	PDI	MIC [μg/mL, (μM)]		HC_{50} [μg/mL, (μM)]	Selectivity	
			E. coli	*B. subtilis*		*E. coli*	*B. subtilis*
Dep-Poly10	10,250	1.07	>500, (>49)	>500, (>49)	>1000, (>98)	-	-
Dep-Poly6	9,950	1.10	200, (20)	300, (30)	>4000, (>400)	>20	>13
Dep-Poly8	10,050	1.13	25, (2.5)	25, (2.5)	<1, (<0.1)	<0.04	<0.04
Dep-Poly9	10,300	1.08	200, (19)	200, (19)	<1, (<0.1)	<0.005	<0.005

This result is consistent with the lack of activity against phospholipid membranes. Introduction of a hydrophobic group at the repeat unit level produced an increase in antibacterial and hemolytic activities, which depended on the size of hydrophobic group. Dep-poly6, with an isopropylidene pendant group, exhibited antibacterial activity with MIC of 200 μg/mL against *E. coli*, which is less efficacious than most antimicrobial peptides, and their mimic, that have MICs typically ranging between 1–50 μg/mL (*23, 25–28, 47, 62–65*). However, the hemolytic activity of dep-poly6 remained below 5% up to 3000 μg/mL, a value well above its MIC. Above 3000 μg/mL the hemolytic activity of this polymer increase more rapidly with increasing concentration. Increase in hemolysis, to 25%, at 4000 μg/mL could be induced through different mechanisms, such as increased osmotic pressure at high polymer concentration, rather than a local membrane perturbation. However HC_{50} value remained above the measured concentration of 4000 µg/mL, thus giving a selectivity, defined as the ratio of HC_{50} to MIC (*49*), greater than 20.

Dep-poly8, with an additional carbon atom per repeat unit, is more hydrophobic than dep-poly6, and has additional mobility of the pendant alkyl group. Dep-poly8 exhibited a substantial increase in antibacterial activity, with MIC of 25 µg/mL for both *E. coli* and *B. subtilis* as well as hemolytic activity, HC_{50} less than 1 µg/mL, with an 80% hemolysis at 1 µg/mL. This increase in antibacterial and hemolytic activity with increasing hydrophobicity is in accordance with literature reports that predict larger hydrophobic groups will have stronger interactions with the inner core of cell membranes, leading to loss of selectivity (*23, 24, 41, 62*). In the case of dep-poly9, when the hydrophobic size was further increased, the hemolytic activity was retained with a 100% hemolysis at 1 µg/mL; however, the antibacterial activity decreased to a MIC of 200 µg/mL. In many instances, hydrophobic interactions have been reported to control hemolytic activities; whereas charge interactions are suggested to be more important for antibacterial activity (*23, 47*). These results show that the presence, and balance, of hydrophobic and hydrophilic groups dictate the antibacterial and hemolytic activities of the amphiphilic, non-natural polymer, in agreement with natural peptide studies. Moreover, copolymers of 6 and 8 generated highly active agents, with MIC of 40 µg/mL and selectivity greater than 100 (*30*).

Table III. Percent hemolysis values at the lower and upper limits of HC_{50} measurements

Polymer	M_n (g/mol)	PDI	% Hemolysis at	
			4000 µg/mL	1 µg/mL
Dep-Poly6	1,600	1.15	24	-
	137,500	1.27	22	-
Dep-Poly8	1,650	1.26	-	78
	57,200	1.70	-	70
Dep-Poly9	5,300	1.09	-	100
	57,000	1.19	-	100

It was previously suggested in the literature that a comparison between new compounds and a reference peptide would be the best indicator for clinical cytotoxicity, and allow a better comparison between different antibacterial agents from different laboratories (*72*). In a control experiment, the activity of a Magainin derivative (MSI-78), a well-known antimicrobial peptide, was measured against the same *E. coli* strain. In comparison to above described homopolymers, MSI-78 exhibited a selectivity of 9.6 that was calculated from an MIC of 12.5 µg/mL and HC_{50} of 120 µg/mL.

Experimental Considerations: Effect of Blood Freshness

All hemolysis results that are reported in this work were obtained using freshly drawn blood from one individual. During the course of this study, the hemolytic activities of polymers were observed to be dependent on the freshness of the blood. Differences were also noted for blood obtained from different individuals. It was determined that blood stored for more than 7 d was more susceptible to hemolysis than freshly drawn blood. These observations were in accordance with previous literature that reported higher susceptibility to hemolysis, caused by a series of cationic antimicrobial peptides, in the case of blood stored for 21 d in 4 °C, as opposed to fresh blood (72). Non-hemolytic polymers, dep-poly6, and dep-poly10 remained non-hemolytic against old blood that was stored for 3 weeks at 4 °C, and blood from different individuals, with HC_{50} values above 4000 µg/mL.

Advantages of ROMP-Based Synthetic Strategy

As mentioned in the introduction section, amphiphilic polymers have attracted attention for a number of biomedical and therapeutic applications. A variety of amphiphilic polymer architecture was considered either as delivery agents for drugs (73–75) and genes (6, 9, 76), structural components in tissue engineering, or active therapeutics, such as antibacterial agents (15–17, 24). A number of reports described the use of ROMP for the preparation of polymers decorated with biologically active agents, including peptides (77), carbohydrates (78), oligonucleotides (79), and anti-cancer drugs (80). These unique materials with high local density of the active groups in the vicinity of the polymer chains warrant further evaluations in therapeutic applications. These synthetic approaches can easily be combined with our approach through copolymerizations, in order to incorporate various polymer segments with distinct, and complementary biological activities. A proper choice of membrane disrupting amphiphilic block, based on modular norbornene derivatives, could provide selective antibacterial activity, as well as facilitate delivery of these multi-component polymeric agents through mammalian cell membranes. With a powerful set of ROMP-based synthetic approaches available, and careful monomer design, the potency of polymeric therapeutic agents can thus be fine-tuned.

Conclusions

Our initial motivation was to develop amenable synthetic approaches for the preparation of amphiphilic polymers with well-controlled structures that would broaden the interface between macromolecular science and biological sciences. Following those efforts, amphiphilic polymers based on modular norbornene derivatives were shown to exhibit good antibacterial activities and high selectivity for bacteria versus red blood cells. Small modifications to the hydrophobic character of the cationic amphiphilic polymer were shown to change dramatically the antibacterial and hemolytic activities. Tuning the hydrophilic/hydrophobic balance and molecular weights of these copolymers allowed preparation of highly selective, antibacterial, non-hemolytic macromolecules. Desired biological activities were maintained across a large range of molecular weights. Furthermore, this study showed the preparation of fully synthetic high molecular weight polymers that mimic the activities of host-defense peptides in the absence of a specific secondary structure. Equally important, we have extended the antimicrobial activity of these designed polymers to include efficient killing against all category A organisms (*B. anthracis*, *F. tularensis*, and *Y. pestis*) as well as other serious pathogens (*L. monocytogenes*, *P. aeruginosa*, and *MRSA*). The overall data strongly suggest this approach will lead to new materials for homeland defense applications, as well as important contributions to the prevention and treatment of bacterial infections.

References

1. Thomas, J. L.; Tirrell, D. A. *Acc. Chem. Res.* **1992**, *25*, 336-342.
2. Stayton, P. S.; Hoffman, A. S.; Murthy, N.; Lackey, C.; Cheung, C.; Tan, P.; Klumb, L. A.; Chilkoti, A.; Wilbur, F. S.; Press, O. W. *J. Controlled Release* **2000**, *65*, 203-220.
3. Kyriakides, T. R.; Cheung, C. Y.; Murthy, N.; Bornstein, P.; Stayton, P. S.; Hoffman, A. S. *J. Controlled Release* **2002**, *78*, 295-303.
4. Gonzalez, H.; Hwang, S. J.; Davis, M. E. *Bioconjugate Chemistry* **1999**, *10*, 1068-1074.
5. Wen, J.; Kim, G. J. A.; Leong, K. W. *J. Controlled Release* **2003**, *92*, 39-48.
6. Hwang, S. J.; Davis, M. E. *Current Opinion in Molecular Therapeutics* **2001**, *3*, 183-191.
7. Cheung, C. Y.; Murthy, N.; Stayton, P. S.; Hoffman, A. S. *Bioconjugate Chemistry* **2001**, *12*, 906-910.

8. Wolfert, M. A.; Dash, P. R.; Nazarova, O.; Oupicky, D.; Seymour, L. W.; Smart, S.; Strohalm, J.; Ulbrich, K. *Bioconjugate Chemistry* **1999**, *10*, 993-1004.
9. Zhao, Z.; Wang, J.; Mao, H. Q.; Leong, K. W. *Advanced Drug Delivery Reviews* **2003**, *55*, 483-499.
10. Wan, A. C. A.; Mao, H. Q.; Wang, S.; Phua, S. H.; Lee, G. P.; Pan, J. S.; Lu, S.; Wang, J.; Leong, K. W. *Journal of Biomedical Materials Research Part B-Applied Biomaterials* **2004**, *70B*, 91-102.
11. Wang, J.; Sun, D. D. N.; Shin-ya, Y.; Leong, K. W. *Macromolecules* **2004**, *37*, 670-672.
12. Wan, A. C. A.; Mao, H. Q.; Wang, S.; Leong, K. W.; Ong, L.; Yu, H. *Biomaterials* **2001**, *22*, 1147-1156.
13. Tashiro, T. *Macromolecular Materials and Engineering* **2001**, *286*, 63-87.
14. Worley, S. D.; Sun, G. *Trends in Polymer Science* **1996**, *4*, 364-370.
15. Stiriba, S. E.; Frey, H.; Haag, R. *Angew. Chem.-Int. Edit.* **2002**, *41*, 1329-1334.
16. Lim, S. H.; Hudson, S. M. *Journal of Macromolecular Science-Polymer Reviews* **2003**, *C43*, 223-269.
17. Thorsteinsson, T.; Loftsson, T.; Masson, M. *Current Medicinal Chemistry* **2003**, *10*, 1129-1136.
18. Kenawy, E. R.; Mahmoud, Y. A. G. *Macromolecular Bioscience* **2003**, *3*, 107-116.
19. Pavlikova, M.; Lacko, I.; Devinsky, F.; Mlynarcik, D. *Collect. Czech. Chem. Commun.* **1995**, *60*, 1213-1228.
20. Tew, G. N.; Liu, D. H.; Chen, B.; Doerksen, R. J.; Kaplan, J.; Carroll, P. J.; Klein, M. L.; DeGrado, W. F. *Proc. Natl. Acad. Sci. U. S. A.* **2002**, *99*, 5110-5114.
21. Gelman, M. A.; Weisblum, B.; Lynn, D. M.; Gellman, S. H. *Organic Letters* **2004**, *6*, 557-560.
22. Arnt, L.; Nusslein, K.; Tew, G. N. *Journal of Polymer Science Part A-Polymer Chemistry* **2004**, *42*, 3860-3864.
23. Andreu, D.; Rivas, L. *Biopolymers* **1998**, *47*, 415-433.
24. van 't Hof, W.; Veerman, E. C. I.; Helmerhorst, E. J.; Amerongen, A. V. N. *Biological Chemistry* **2001**, *382*, 597-619.
25. Porter, E. A.; Wang, X.; Lee, H. S.; Weisblum, B.; Gellman, S. H. *Nature* **2000**, *405*, 298-298.
26. Fernandez-Lopez, S.; Kim, H. S.; Choi, E. C.; Delgado, M.; Granja, J. R.; Khasanov, A.; Kraehenbuehl, K.; Long, G.; Weinberger, D. A.; Wilcoxen, K. M.; Ghadiri, M. R. *Nature* **2001**, *412*, 452-455.
27. Liu, D. H.; DeGrado, W. F. *J. Am. Chem. Soc.* **2001**, *123*, 7553-7559.
28. Patch, J. A.; Barron, A. E. *J. Am. Chem. Soc.* **2003**, *125*, 12092-12093.
29. Ilker, M. F.; Schule, H.; Coughlin, E. B. *Macromolecules* **2004**, *37*, 694-700.

30. Ilker, M. F.; Nusslein, K.; Tew, G. N.; Coughlin, E. B. *J. Am. Chem. Soc.* **2004**, *126*, 15870-15875.
31. Ivin, K. J., Mol, J. C. *Olefin Metathesis and Metathesis Polymerization*; Academic Press: San Diego, CA, 1997.
32. Stone, K. J.; Little, R. D. *J. Org. Chem.* **1984**, *49*, 1849-1853.
33. Choi, T. L.; Grubbs, R. H. *Angew. Chem.-Int. Edit.* **2003**, *42*, 1743-1746.
34. Oren, Z.; Shai, Y. *Biopolymers* **1998**, *47*, 451-463.
35. Menger, F. M.; Seredyuk, V. A.; Kitaeva, M. V.; Yaroslavov, A. A.; Melik-Nubarov, N. S. *J. Am. Chem. Soc.* **2003**, *125*, 2846-2847.
36. Murthy, N.; Robichaud, J. R.; Tirrell, D. A.; Stayton, P. S.; Hoffman, A. S. *J. Controlled Release* **1999**, *61*, 137-143.
37. Helander, I. M.; Latva-Kala, K.; Lounatmaa, K. *Microbiology* **1998**, *144*, 385-390.
38. Arnt, L.; Tew, G. N. *J. Am. Chem. Soc.* **2002**, *124*, 7664-7665.
39. Yaroslavov, A. A.; Yaroslavova, E. G.; Rakhnyanskaya, A. A.; Menger, F. M.; Kabanov, V. A. *Colloids and Surfaces B-Biointerfaces* **1999**, *16*, 29-43.
40. Oku, N.; Yamaguchi, N.; Shibamoto, S.; Ito, F.; Nango, M. *Journal of Biochemistry* **1986**, *100*, 935-944.
41. Zasloff, M. *Nature* **2002**, *415*, 389-395.
42. Hamuro, Y.; Schneider, J. P.; DeGrado, W. F. *J. Am. Chem. Soc.* **1999**, *121*, 12200-12201.
43. Porter, E. A.; Wang, X. F.; Lee, H. S.; Weisblum, B.; Gellman, S. H. *Nature* **2000**, *404*, 565-565.
44. Oren, Z.; Shai, Y. *Biochemistry* **1997**, *36*, 1826-1835.
45. Hancock, R. E. W.; Chapple, D. S. *Antimicrob. Agents Chemother.* **1999**, *43*, 1317-1323.
46. Huang, H. W. *Biochemistry* **2000**, *39*, 8347-8352.
47. Dathe, M.; Schumann, M.; Wieprecht, T.; Winkler, A.; Beyermann, M.; Krause, E.; Matsuzaki, K.; Murase, O.; Bienert, M. *Biochemistry* **1996**, *35*, 12612-12622.
48. Tytler, E. M.; Anantharamaiah, G. M.; Walker, D. E.; Mishra, V. K.; Palgunachari, M. N.; Segrest, J. P. *Biochemistry* **1995**, *34*, 4393-4401.
49. Liu, D. H.; Choi, S.; Chen, B.; Doerksen, R. J.; Clements, D. J.; Winkler, J. D.; Klein, M. L.; DeGrado, W. F. *Angew. Chem.-Int. Edit.* **2004**, *43*, 1158-1162.
50. Zhao, H. X.; Bose, S.; Tuominen, E. K. J.; Kinnunen, P. K. J. *Biochemistry* **2004**, *43*, 10192-10202.
51. Pokorny, A.; Almeida, P. F. F. *Biochemistry* **2004**, *43*, 8846-8857.
52. Juvvadi, P.; Vunnam, S.; Merrifield, R. B. *J. Am. Chem. Soc.* **1996**, *118*, 8989-8997.
53. Almeida, P. F. F.; Vaz, W. L. C.; Thompson, T. E. *Biochemistry* **1992**, *31*, 6739-6747.

54. Hotani, H.; Nomura, F.; Suzuki, Y. *Current Opinion in Colloid & Interface Science* **1999**, *4*, 358-368.
55. Kisak, E. T.; Coldren, B.; Evans, C. A.; Boyer, C.; Zasadzinski, J. A. *Current Medicinal Chemistry* **2004**, *11*, 199-219.
56. Drummond, D. C.; Zignani, M.; Leroux, J. C. *Progress in Lipid Research* **2000**, *39*, 409-460.
57. Noble, C. O.; Kirpotin, D. B.; Hayes, M. E.; Mamot, C.; Hong, K.; Park, J. W.; Benz, C. C.; Marks, J. D.; Drummond, D. C. *Expert Opinion on Therapeutic Targets* **2004**, *8*, 335-353.
58. Santos, N. C.; Castanho, M. *Quimica Nova* **2002**, *25*, 1181-1185.
59. Holopainen, J. M.; Angelova, M.; Kinnunen, P. K. J. *Liposomes, Pt A* **2003**, *367*, 15-23.
60. Antonietti, M.; Forster, S. *Advanced Materials* **2003**, *15*, 1323-1333.
61. Vaz, W. L. C.; Almeida, P. F. F. *Current Opinion in Structural Biology* **1993**, *3*, 482-488.
62. Hancock, R. E. W. *Drugs* **1999**, *57*, 469-473.
63. Wade, D.; Boman, A.; Wahlin, B.; Drain, C. M.; Andreu, D.; Boman, H. G.; Merrifield, R. B. *Proc. Natl. Acad. Sci. U. S. A.* **1990**, *87*, 4761-4765.
64. Raguse, T. L.; Porter, E. A.; Weisblum, B.; Gellman, S. H. *J. Am. Chem. Soc.* **2002**, *124*, 12774-12785.
65. Schmitt, M. A.; Weisblum, B.; Gellman, S. H. *J. Am. Chem. Soc.* **2004**, *126*, 6848-6849.
66. Li, G. J.; Shen, J. R.; Zhu, Y. L. *J. Appl. Polym. Sci.* **1998**, *67*, 1761-1768.
67. Rowden, A.; Cutarelli, P. E.; Cavanaugh, T. B.; Sellner, P. A. *Investigative Ophthalmology & Visual Science* **1997**, *38*, 5135-5135.
68. Liu, N. H.; Khong, D.; Chung, S. K.; Hwang, D. G. *Investigative Ophthalmology & Visual Science* **1996**, *37*, 4058-4058.
69. Vogelberg, K.; Boehnke, M. *Investigative Ophthalmology & Visual Science* **1994**, *35*, 1337-1337.
70. Albert, M.; Feiertag, P.; Hayn, G.; Saf, R.; Honig, H. *Biomacromolecules* **2003**, *4*, 1811-1817.
71. Messick, C. R.; Pendland, S. L.; Moshirfar, M.; Fiscella, R. G.; Losnedahl, K. J.; Schriever, C. A.; Schreckenberger, P. C. *J. Antimicrob. Chemother.* **1999**, *44*, 297-298.
72. Helmerhorst, E. J.; Reijnders, I. M.; van't Hof, W.; Veerman, E. C. I.; Amerongen, A. V. N. *FEBS Lett.* **1999**, *449*, 105-110.
73. D'Souza, A. J. M.; Topp, E. M. *J. Pharm. Sci.* **2004**, *93*, 1962-1979.
74. Pawar, R.; Ben-Ari, A.; Domb, A. J. *Expert Opinion on Biological Therapy* **2004**, *4*, 1203-1212.
75. Ghosh, S. *Journal of Chemical Research-S* **2004**, 241-246.
76. Pannier, A. K.; Shea, L. D. *Molecular Therapy* **2004**, *10*, 19-26.
77. Maynard, H. D.; Okada, S. Y.; Grubbs, R. H. *Macromolecules* **2000**, *33*, 6239-6248.

78. Mortell, K. H.; Gingras, M.; Kiessling, L. L. *J. Am. Chem. Soc.* **1994**, *116*, 12053-12054.
79. Watson, K. J.; Park, S. J.; Im, J. H.; Nguyen, S. T.; Mirkin, C. A. *J. Am. Chem. Soc.* **2001**, *123*, 5592-5593.
80. Watson, K. J.; Anderson, D. R.; Nguyen, S. T. *Macromolecules* **2001**, *34*, 3507-3509.

Chapter 12

Catalysts for Aerobic Decontamination of Chemical Warfare Agents under Ambient Conditions

Craig L. Hill[1], Nelya M. Okun[1,2], Daniel A. Hillesheim[1], and Yurii V. Geletii[1]

[1]Department of Chemistry, Emory University, Atlanta, GA 30122
[2]Current address: Chemence Medical, 185 Bluegrass Valley Parkway, Alpharetta, GA 30005

Transition-metal-substituted polyoxometalates (POMs) in the presence of NO_x species, based on studies with the mustard (HD) simulant 2-chloroethyl ethyl sulfide (CEES), are the most reactive catalysts yet for the desired aerobic (O_2/air-based) decontamination of mustard (fast and quantitatively selective for sulfoxide using the ambient environment—air at room temperature). The properties and catalytic CEES sulfoxidation chemistry of one complex, $Fe^{III}[H(ONO_2)_2]PW_{11}O_{39}^{5-} \cdot HNO_3$ (**1**) are addressed. Product distribution, structural and mechanistic studies on these highly complex catalytic systems are presented. In addition, preliminary results on CEES sulfoxidation catalyzed by $Ag_2[(CH_3C(CH_2O)_3)_2V_6O_{13}]$, a model for repeating units in a range of POM-based catalytically decontaminating and/or detecting materials, are described.

Introduction

Two of the four chemical warfare agents (CWAs) of great concern, mustard (bis(2-chloroethyl)sulfide or HD) and VX, are amenable to decontamination by oxidative means, as well as hydrolytic and other means (*1*). Ideal oxidative decontamination catalysts would convert CWAs to nontoxic products rapidly and selectively, at ambient temperature, using only O_2 (air) as the oxidant, and would have sufficient stability to be used for long periods of time and/or deployed repeatedly. Catalytic decontamination systems have two marked and general advantages over stoichiometric ones: a far smaller quantity of the decontaminating compound ("active") is needed and this "active" compound is not consumed in the decontamination process itself. These attributes could, in principle, lead to decontaminating and protective materials (fabrics, coatings, cosmetics, filters, etc.) that are durable, or even multiply re-useable. A plethora of catalysts for oxidation of organic substrates similar to the principal CWAs by peroxides and several other oxidants heavily used in industry, including the green oxidant, H_2O_2, are readily available and under continuing development. However, catalysts for O_2-based oxidations that are rapid and selective are far fewer in number. This derives from two general characteristics associated with O_2 reactions. First, such reactions typically proceed either very slowly (days-years, e.g., degradation of most plastics, foods and other consumer materials) or very rapidly under autogenous conditions (seconds or less, e.g., rapid heating from substantial exothermicities resulting in combustion). Second, O_2-based reactions, including those that proceed at low temperature, are nearly all radical chain processes (autoxidations). These transformations, by virtue of their complexity (multiple and branching chain steps and frequent autocatalytic components), are intrinsically low in selectivity. Furthermore, mustard is a sulfide and sulfides are not oxidized by autoxidation at the low temperatures needed in viable protection/decontamination technologies.

The oxidative decontamination process for mustard, a CWA threat of major concern, targeted by us and others, is selective sulfoxidation, eq 1, because the sulfoxide has been known for decades to be far less toxic than mustard itself (*2, 3*).

$$\text{mustard (HD)} + 0.5\ O_2 \longrightarrow \text{mustard sulfoxide (HDO)} \quad (1)$$

mustard (HD)
(toxic vesicant)

mustard sulfoxide (HDO)
(far less toxic than HD)

In contrast, the corresponding sulfone, a product of over-oxidation, has substantial toxicity (fairly potent vesicant). Thus, viable oxidative decontamination systems must exhibit high selectivity for sulfoxide as well as high rates.

Recently, several groups have reported catalysts for the O_2-based oxidation of sulfides that are both fairly fast and selective. Bosch and Kochi established that NO_2 is a highly effective sulfoxidation catalyst (4, 5), as is NO; however, gases are not compatible with the formulation of useful self-decontaminating solutions, fabrics, cosmetics, filters or coatings. Other published catalysts for aerobic sulfoxidation are corrosive, toxic, require light, or exhibit other characteristics we wish to avoid.

Our group, in collaboration with the groups of H. Schreuder-Gibson, J. Walker and D. Dubose at the Natick Soldier Center, E. Braue and co-workers at USAMRICD, and others in national laboratories and small companies with support from the Army Research Office, have developed several catalysts for eq 1. The most effective thus far have been based on polyoxometalates (POMs) and/or coinage metal (Cu, Ag, Au) complexes. POMs are attractive in that their elemental compositions and structures, which influence or control reduction potentials, solubilities, shapes, charges/polarities, and other molecular parameters impacting utility, can be extensively varied (6-8). In addition, syntheses of POMs can generally be realized that are green and amenable to scale-up. POMs are effectively non-toxic (9) and recent tests in collaboration with our co-workers at Natick and an independent evaluation company indicated that our most active catalysts (addressed below in this chapter) have no apparent toxicity or irritation tendencies when in direct contact with skin (unpublished results) (10).

In 2001, $Ag_5PV_2Mo_{10}O_{40}$ was structurally, spectroscopically and chemically characterized and reported to catalyze eq 1 (11). Although it is highly selective for the sulfoxide and functions as a solid under ambient conditions, it is too slow to be of practical interest. Also in 2001, Au(III)(NO_3)(halide)(sulfide) complexes that also catalyze eq 1 selectively, using ambient air, were developed and well characterized (12). However, these systems are unusually complex and their activities are not very reproducible. Furthermore, gold may be prohibitively expensive for large scale use unless turnover rates and stabilities are very high. In 2003, a coordination polymer between Cu(II) and $PV_2Mo_{10}O_{40}^{5-}$ that catalyzed eq 1 under ambient conditions was structurally and spectroscopically characterized (13). In 2003 and 2004, robust heterogeneous catalysts comprised of negatively charged POMs electrostatically bonded to cationic silica nanoparticles were reported (14). These catalyze several oxidations, including eq 1, selectively and fairly rapidly using air at 25 °C.

We report here a new catalyst for eq 1 (using the mustard simulant 2-chloroethyl ethyl sulfide, CEES, to the corresponding sulfoxide, CEESO) that is more active than any of the above systems. It is comprised of an iron-substituted POM (specifically an alpha-Keggin complex) in the presence of nitrate and H^+. Elemental analyses, vibrational spectra and other data are consistent with the presence of a hydrogen dinitrate group and an additional nitrate of crystallization: $Fe^{III}[H(ONO_2)_2]PW_{11}O_{39}^{5-} \cdot HNO_3$ (1). We also report preliminary work on the sulfoxidation of CEES catalyzed by

$Ag_2[(CH_3C(CH_2O)_3)_2V_6O_{13}]$ (2). This POM contains two hydrolytically robust triester groups. These bis(triester)V_6 units are readily incorporated into a range of POM-based materials that are targeted to both detect (by color change) and catalytically decontaminate oxidatively-prone CWAs and several environmental pollutants.

$Fe^{III}[H(ONO_2)_2]PW_{11}O_{39}^{5-} \cdot HNO_3$ (1), a highly effective catalyst for aerobic sulfoxidation

The highly-reactive, aerobic sulfoxidation catalyst, **1**, is prepared from alpha-$PW_{11}O_{39}^{7-}$ and $Fe(NO_3)_3$ in acetonitrile at ambient temperature and isolated as a mixed tetra-*n*-butylammonium (TBA)/H^+ salt (typically a TBA_3H_2 salt) (*15*). Unfortunately, the unequivocal structural characterization of **1** is very challenging because the two most powerful appropriate techniques, NMR (multinuclear and multidimensional) and single crystal X-ray diffraction, are not informative. The paramagnetism of **1** renders NMR peak widths too broad to be useful and four attempted X-ray structure determinations of mixed TBA salts of **1** all failed for the same predictable reason: all mono-substituted alpha-Keggin ($XM_{12}O_{40}^{n-}$) complexes (~T_d point group symmetry) are statistically twelve-fold disordered. This renders identification of the terminal ligands, the NO_x units in the case of **1**, via X-ray analysis effectively impossible. Characterization via FTIR spectrum shows the presence of typical bands for the alpha-$PW_{11}O_{39}$ (1022 (w), 965 (s), 891 (m), 815 (s), 668 (w), 591 (w) and 515 (m) cm^{-1}) and hydrogen dinitrate groups (1383, 1481, and 1665 cm^{-1}) (*15*). Extensive comparisons of the infrared spectra of NO_x species including NO^+, NO_2 (g), bridging, monodentate ($M-O-NO_2$) or bidentate ($M-O_2-NO$) nitrate, and unligated (C_{2v} NO_2^-) or monodentate ($M-O-NO$) nitrite with the infrared spectrum of **1** rule out the presence of these NO_x moieties in **1**. The authentic iron-nitrosyl complex, $TBA_6Fe^{II}(NO)PW_{11}O_{39}$ (*16*), was also synthesized, and this compound, along with most of the above NO_x species, were shown to be inactive for eq 1 in the absence of protons, under ambient conditions. Complex **1** was also subjected to analysis by TGA and DSC, which indicated loss of the NO_x moieties at ~105 and ~200 °C. Elemental analyses and wet chemical analyses were also consistent with the formulation of **1** given above: $Fe^{III}[H(ONO_2)_2]PW_{11}O_{39}^{5-} \cdot HNO_3$.

In order to precisely define the structure of the NO_x units in **1**, future efforts will involve the X-ray crystallographic characterization mono-substituted beta-Keggin isomers. These isomers, unlike the mono-substituted alpha-Keggin unit in **1**, generally produce single crystals in which POM moiety is not positionally disordered.

Table I compares the reactivity of **1** with that of other catalysts for eq 1 (using CEES) under ambient conditions (oxidant = air; room temperature). All these reactions have very high, or effectively quantitative, selectivities for the desired sulfoxide. Furthermore, the reactions catalyzed by 1 go to 100% conversion readily, a point of practical consequence in a theatre of operations.

A number of control reactions were carried out in order to assess the importance of the individual components of $TBA_3H_2\mathbf{1}$. First, a model (Table I, entry 2) of the active catalyst comprising $TBA_4Fe(H_2O)PW_{11}O_{39}$, $TBANO_3$ and a proton source (p-toluenesulfonic acid, pTsOH) in the same mole ratios to entry 1 was evaluated. The catalytic activity of this model system is similar to that of the active catalyst. Upon removal of the POM from the system, the catalytic activity is lower (entry 3). The addition of a proton-specific base, 2,6-di-t-butylpyridine, to the active catalyst, $TBA_3H_2\mathbf{1}$, results in complete loss of activity (entry 4). Other NO_x species (including, but not limited to, entries 5 and 6) show minimal activity. Thus, all components of the reactive system, POM, nitrate, and proton, are essential for optimal catalysis of eq 1.

Table I. Homogeneous Air-Based Oxidation of 2-Chloroethyl Ethyl Sulfide (CEES) by the Most Active Known Catalysts [a]

	Catalyst	[Catalyst], (mM)	Conversion, %[b]	TON[c]
1	$TBA_3H_2\mathbf{1}$[d]	2.0	92.8	163
2	$TBA_4Fe(H_2O)PW_{11}O_{39}$[e]	2.0	~99	174
3	$TBANO_3$[f]	6.0	6	10.5
4	$TBA_3H_2\mathbf{1}$[g]	5.0	0	0
5	$NOPF_6$	2.0	2.8	3.5
6	NO_2BF_4	2.0	2.5	4.4
7	$(NH_4)_2Ce^{IV}(NO_3)_6$	2.4	28.0	49
8	$Ti(NO_3)_4$	3.2	29.5	52

[a] General conditions: 0.35 M of CEES, catalyst (given in column 2), 1 atm of air, 1,3-dichlorobenzene (internal standard) in 2.3 mL of acetonitrile at 25 °C for 40 h in a 20-mL vial. [b] conversion = (mol of CEES consumed/mol of initial CEES) x 100; [c] turnover number = (moles of CEESO/moles of catalyst); [d] $\mathbf{1} = [Fe^{III}[H(ONO_2)_2]PW_{11}O_{39}]^{5-}\cdot HNO_3$. [e] In the presence of 6.0 mM of $TBANO_3$ and 8.0 mM of pTsOH (equivalent to concentrations in $TBA_3H_2\mathbf{1}$). [f] In the presence of 8.0 mM of pTsOH (equivalent to concentrations in $TBA_3H_2\mathbf{1}$). [g] In the presence of 20 mM 2,6-di-t-butylpyridine.

The activities of other nitrate-containing sulfoxidation catalysts were also evaluated. The known sulfide oxidation catalysts $(NH_4)_2Ce^{IV}(NO_3)_6$ (17) (entry 7) and titanium nitrate (entry 8) have moderate activity. The data demonstrate the essential role of both nitrate and proton in the catalytic, aerobic oxidation of the sulfide CEES.

Catalysis by Nitrate / Proton System

In order to clarify the role of nitrate, proton and POM and their possible synergism in the catalytic aerobic oxidation of CEES, we attempted to simplify further the reference system (entry 1 in Table I). Common synthetic routes to organic soluble POMs often include the use of TBABr. As such, we explored the possible effects of residual bromide on the catalytic activity. $TBANO_3$, TBABr and pTsOH were used as sources of nitrate, bromide, and proton, respectively, while $TBA_4Fe(H_2O)PW_{11}O_{39}$ was used as a Fe-containing POM relevant to **1**. Furthermore, to address a role of the central heteroatom in Keggin-type POMs, $TBA_4Fe(H_2O)PW_{11}O_{39}$ was replaced with $TBA_5Fe(H_2O)SiW_{11}O_{39}$. Then to address a role of the addendum transition metal, $TBA_5Fe(H_2O)SiW_{11}O_{39}$ was replaced with $TBA_6Cu(H_2O)SiW_{11}O_{39}$. As a control, the Cu-POM was replaced with $Cu(ClO_4)_2$. Additionally, we altered experimental conditions to increase the reaction rates by raising catalyst and dioxygen concentrations and temperature.

The catalytic system consisting only of nitrate and proton shows limited activity under both the conditions detailed in Table I (entry 3) and the minimally

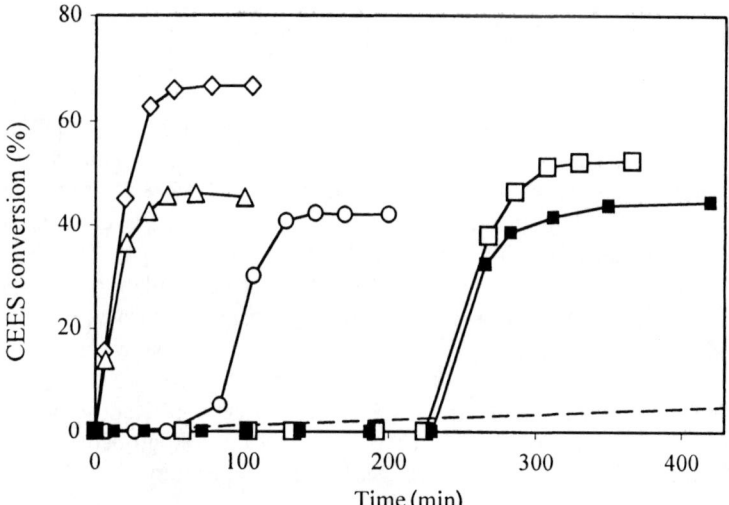

Figure 1. Oxidation of CEES (0.5 M) to CEESO catalyzed by nitrate (8 mM $TBANO_3$)/proton (42 mM pTsOH)/bromide (8 mM TBABr) systems in acetonitrile at 30 °C and 1 atm O_2. The reactions when the following additional components are added to the reference catalytic system above (○) are shown: (■) plus 3.6 mM $TBA_4Fe(H_2O)PW_{11}O_{39}$, (□) plus $TBA_5Fe(H_2O)SiW_{11}O_{39}$, (◇) plus $TBA_6Cu(H_2O)SiW_{11}O_{39}$, (△) plus $CuClO_4$, (dashed line) reference system but no bromide, no POM.

optimized conditions in Figure 1 (dashed line). In addition, bromide functions as a co-catalyst in the aerobic oxidation of CEES by this simplified system. In the presence of TBABr, the reaction proceeds ca. two orders of magnitude faster, but in a more complicated manner. There is a significant induction period and the reaction terminates before reaching complete conversion (Figure 1). CEESO is the only detectable product, as in the reference system. Addition of $TBA_4Fe(H_2O)PW_{11}O_{39}$ to the nitrate/bromide/proton system does not affect the rate, or the yield, but considerably increases the induction period of the reaction. Substitution of $TBA_5Fe(H_2O)SiW_{11}O_{39}$ for $TBA_4Fe(H_2O)PW_{11}O_{39}$ results in a modest, but reproducible, increase in yield. Significantly, replacement of $TBA_4Fe(H_2O)PW_{11}O_{39}$ with $TBA_6Cu(H_2O)SiW_{11}O_{39}$ completely eliminates the induction period and increases both the reaction rate and product yield. Further simplification of the system by replacing the Cu-POM with $Cu(ClO_4)_2$ only removes induction period; the yield remains unchanged. These control studies indicate that the best catalysts for eq 1 will likely contain Cu-containing POMs, bromide, nitrate and acid. It should be noted that bromide and nitrate are very benign (found in many foods and cosmetics). Likewise, groups more acidic than **1** are found in many materials that contact human skin.

Catalysis by $Ag_2[(CH_3C(CH_2O)_3)_2V_6O_{13}]$ (2)

$Ag_2[(CH_3C(CH_2O)_3)_2V_6O_{13}]$ (**2**) is one of many possible salts of compounds or materials that contain one or more bis(triester)V_6 units. Compound **2** is prepared by procedures similar to those in the literature (*18*) and as described below in the Experimental section. Initially the TBA salt is prepared; a metathesis reaction then provides the Ag_2 salt. Bis(triester)V_6 units are known to catalyze the aerobic oxidation of thiols (2 RSH + 1/2 O_2 → RSSR + H_2O) under ambient conditions (unpublished results) but sulfoxidation had not been assessed prior to this work. Preliminary results indicate that aerobic sulfoxidation of CEES catalyzed by **2** is probably too slow under ambient conditions to be of interest. While other salts of $[(CH_3C(CH_2O)_3)_2V_6O_{13}]^{2-}$ including H^+ or mixed metal/H^+ salts are likely to be significantly more active, we chose to assess the oxidation of CEES by *tert*-butyl hydroperoxide (TBHP) because this oxidant is inexpensive, readily available, transportable and quite safe.

Sulfoxidation by TBHP catalyzed by **2**, eq 2, shows an induction period. However, once eq 2 starts, it proceeds rapidly at 25 °C to completion.

Representative kinetics are shown in Figure 2. The reaction can be followed by ^{51}V NMR, electronic absorption spectroscopy, and gas chromatography.

$$\text{CEES + TBHP} \rightarrow \text{CEESO + }t\text{-butyl alcohol (TBA)} \qquad (2)$$

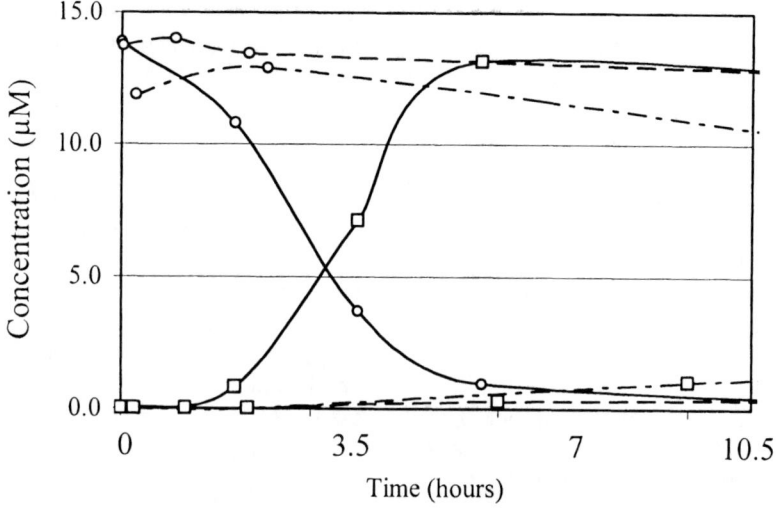

Figure 2. CEES ( ) oxidation to CEESO ( ) by 14.8 mM TBHP in acetonitrile at room temperature. Solid lines represent oxidation in the presence of 3.3 mM $Ag_2[(CH_3C(CH_2O)_3)_2V_6O_{13}]$ (2). Dashed lines represent the uncatalyzed background reaction. The dash-dot lines represent the background oxidation in the presence of 6.7 mM $AgClO_4$.

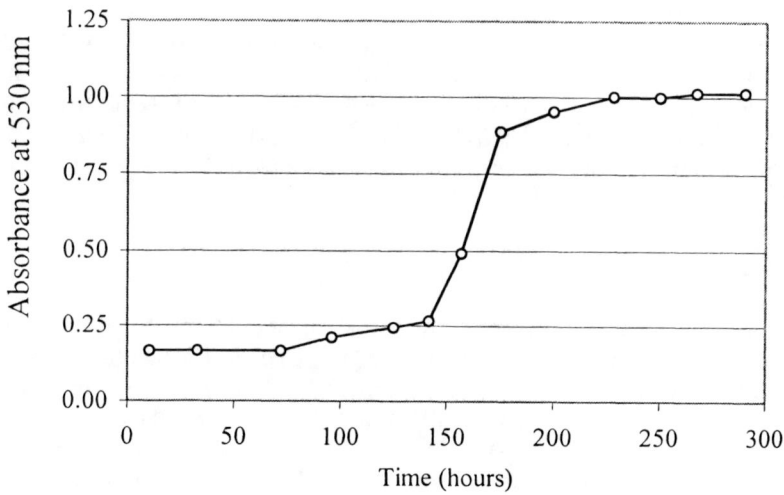

Figure 3. UV-vis absorbance at 530 nm. The oxidation of CEES to CEESO by TBHP catalyzed by 2 was monitored by UV-vis spectroscopy. Changes in the UV-vis spectra correspond roughly to similar changes in the catalyzed reactions themselves. Reaction conditions are described in Figure 2.

The single singlet peak in the ^{51}V NMR spectrum of **2** at the outset of the reaction (all 6 vanadium centers equivalent by symmetry) decreases and multiple signals increase. These changes closely parallel the activity. The electronic absorption spectra also exhibit a time dependence that parallels the onset of catalytic activity (Figure 3). These changes in the ^{51}V NMR and UV-visible spectra are not seen if any one of the components of the reaction is missing. That is, incubating **2** in acetonitrile with each of the individual components (sulfide or oxidant or solvent) affords no change in spectra or any evidence of the oxidation product (CEESO).

The spectral changes are consistent with, but do not prove the formation of, a transient alkylhydroperoxide derivative of, the V_6 cage (replacement of a μ-oxo bridge with an alkyl hydroperoxo group), which might be the catalytic species. While only 4 turnovers are observed, the reaction is evidently catalytic, as the reaction does not proceed in the absence of catalyst on the same time scale. Efforts are underway to isolate and fully characterize this lower-symmetry vanadium-containing specie(s).

Experimental

Materials

All chemicals were commercially-available, reagent grade and used as received from commercial sources, unless otherwise specified. The isopolyoxovanadates [n-(C$_4$H$_9$)$_4$N]$_3$[H$_3$V$_{10}$O$_{28}$](*19*), [n-(C$_4$H$_9$)$_4$N]$_3$[V$_5$O$_{14}$](*20*), and [(n-(C$_4$H$_9$)$_4$N]$_2$[(CH$_3$C(CH$_2$O)$_3$)$_2$V$_6$O$_{13}$](*21*) were prepared by literature methods. The polyoxometalates Fe(H$_2$O)PW$_{11}$O$_{39}$$^{4-}$, Fe(H$_2$O)SiW$_{11}O_{39}$$^{5-}$, Cu(H$_2$O)SiW$_{11}O_{39}$$^{6-}$, and Fe^{III}[H(ONO$_2$)$_2$]PW$_{11}O_{39}$$^{5-}$·HNO$_3$ were synthesized as tetrabutylammonium salts as per literature methods (*15, 22, 23*).

Instrumentation

Elemental analysis was performed by Atlantic Microlabs, Atlanta, GA (carbon, hydrogen, and nitrogen) and Kontilabs, Quebec, Canada (all elements). ^{51}V NMR spectra were acquired on a Varian 600 MHz Unity spectrometer and externally referenced to 10 mM H$_4$[PVMo$_{11}$O$_{40}$] in 0.6 M NaCl (∂ = -533.6 ppm relative to neat VOCl$_3$) (*24*). Solution ^{1}H and ^{13}C NMR spectroscopic measurements were made on a Varian INOVA 400 MHz spectrometer and resonance signals were referenced to residual solvent signals. Infrared spectra (3–5 wt % in KBr) were recorded on a Thermo Electron Corporation Nicolet 6700 FTIR spectrometer. Catalytic reactions (reactant and product) were quantified using a Hewlett-Packard 5890 GC equipped with a HP-5 poly (5%

diphenyl/95% dimethyl) siloxane capillary column and an FID detector. UV-vis spectra of the materials and reactions were acquired using a Hewlett-Packard 8452A diode array spectrophotometer.

Reactions Catalyzed by 1 and the NO_3^-/Proton Model System

Specific conditions for these reactions have been given in the figure captions or in the text. Typically, TBANO$_3$ (8 mM), pTsOH (42 mM), TBABr (8 mM) and internal standard were dissolved in acetonitrile (2.3 mL) under 1 atm O$_2$. After temperature equilibration, the reaction was initiated by addition of CEES (0.5 M). Catalytic reactions, as well as blanks and controls, were run, in parallel, at room temperature or at 30 °C, with stirring at 300 rpm, in septum-top, scintillation vials. In select cases, a capped, quartz cuvette equipped with a magnetic stirrer was used to obtain UV-vis spectra of the compounds and reactions. Reactions were monitored by gas chromatography at defined time intervals and concentrations of CEES and CEESO were calculated using internal standard techniques.

Synthesis of $Ag_2[(CH_3C(CH_2O)_3)_2V_6O_{13}]$ (2)

To a solution of $[(n-(C_4H_9)_4N]_2[(CH_3C(CH_2O)_3)_2V_6O_{13}]$ (0.121 g) in acetonitrile (6 mL) is added an excess amount of NaClO$_4$ (0.5 g). A precipitate formed upon shaking and was isolated via filtration and dried in vacuo. The precipitate was redissolved in water and three molar equivalents of AgNO$_3$ were added. The resulting red-orange precipitate was separated by filtration, washed with cold water, then ether, then dried in vacuo. Pure crystals were obtained by ether diffusion into a concentrated acetonitrile solution. ^{1}H NMR (400 MHz, CD$_3$CN): ∂ = 4.897 (s), 0.767 (s). ^{51}V NMR (600 MHz, CD$_3$CN): ∂ = -498.494 (s). FT-IR (Diamond ATR): 2955, 2923, 1614 (weak broad), 1452, 1394, 1369 (weak), 1204 (weak), 1122, 1013 (very strong), 967 (very strong), 952 (very strong), 932 (shoulder), 784 (broad), 684 (broad) cm^{-1}.

Reactions Catalyzed by 2

A solution of CEES (13.7 mM), 1,3-dichlorobenzene (internal standard, 14 mM), and TBHP (14.8 mM) in acetonitrile was prepared. Three septum-capped, scintillation vials were equipped with magnetic stirrers and charged with 3 mL each of the prepared solution. Nothing else was added to vial 1. To vial 2 were added 10 mg (0.010 mmol) of Ag$_2$[(CH$_3$C(CH$_2$O)$_3$)$_2$V$_6$O$_{13}$], **2**. To vial 3 were added 4 mg AgClO$_4$ (0.020 mmol). The mixtures were stirred at 300 rpm, at room temperature. Reactions were monitored via GC and concentrations of CEES and CEESO were calculated using internal standard techniques.

Acknowledgments

We thank the U.S. Army Research Office (ARO; grants DAAD19-01-1-0593 and current W911NF-05-1-0200) and the Natick Soldier Center (grant DAAD19-02-D-0001) for support of this research.

References

1. Yang, Y. C.; Baker, J. A.; Ward, J. R. *Chem. Rev.* **1992**, *92*, 1729-1743.
2. Hirade, J.; Ninomiya, A. *J. Biochem.* (Tokyo) **1950**, *37*, 19-26.
3. Anslow, J.; W. P.; Karnofsky, D. A.; Val Jager, B.; Smith, H. W. *J. Pharmacol. Exp. Ther.* **1948**, *93*, 1-9.
4. Bosch, E.; Kochi, J. K. *J. Org. Chem.* **1995**, *60*, 3172-3183.
5. Bosch, E.; Kochi, J. K. In *N-Centered Radicals*; Alfassi, Z. B., Ed.; John Wiley & Sons: Chichester, 1998, pp 68-128.
6. *Polyoxometalate Chemistry From Topology via Self-Assembly to Applications*; Pope, M. T.; Müller, A., Eds.; Kluwer Academic Publishers: Dordrecht, 2001.
7. *Polyoxometalate Chemistry for Nano-Composite Design*; Yamase, T.; Pope, M. T., Eds.; Kluwer Academic/Plenum Publishers: New York, 2002; Vol. 2.
8. Hill, C. L. In *Comprehensive Coordination Chemistry-II: From Biology to Nanotechnology*; Wedd, A. G., Ed.; Elsevier: Oxford, 2004; Vol. 4, pp 679-759.
9. Rhule, J. T.; Hill, C. L.; Judd, D. A.; Schinazi, R. F. *Chem. Rev.* **1998**, *98*, 327-357.
10. DuBose, D. A.; Blaha, M. D.; Morehouse, D. H.; Okun, N.; Hill, C. L. *J. Medical Chemical Defense* **2004**, *2*, 1-12.
11. Rhule, J. T.; Neiwert, W. A.; Hardcastle, K. I.; Do, B. T.; Hill, C. L. *J. Am. Chem. Soc.* **2001**, *123*, 12101-12102.
12. (a) Boring, E.; Geletii, Yu.; Hill, C. L. *J. Am. Chem. Soc.* **2001**, *123*, 1625-1635; (b) Boring, E.; Geletii, Yu.; Hill, C. L. *J. Mol. Cat. A* **2001**, *176*, 49-63.
13. Okun, N. M.; Anderson, T. M.; Hardcastle, K. I.; Hill, C. L. *Inorg. Chem.* **2003**, *42*, 6610-6612.
14. (a) Okun, N. M.; Anderson, T. M.; Hill, C. L. *J. Am. Chem. Soc.* **2003**, *125*, 3194-3195; (b) Okun, N. M.; Ritorto, M. D.; Anderson, T. M.; Apkarian, R. P.; Hill, C. L. *Chem. Mater.* **2004**, *16*, 2551-2558.
15. Okun, N.; Tarr, J.; Hillesheim, D.; Zhang, L.; Hardcastle, K.; Hill, C. L. *J. Mol. Catal. A* **2006**, *246*, 11-17.
16. Toth, J. E.; Anson, F. C. *J. Am. Chem. Soc.* **1989**, *111*, 2444-2451.
17. Riley, D. P.; Smith, M. R.; Correa, P. E. *J. Am. Chem. Soc.* **1988**, *110*, 177-180.

18. (a) Chen, Q.; Goshorn, D. P.; Scholes, C. P.; Tan, X. L.; Zubieta, J. *J. Am. Chem. Soc.* **1992**, *114*, 4667-4681; (b) Neiwert, W. A. Ph.D. Thesis, Emory University, 2004.
19. Day, V. W.; Klemperer, W. G.; Maltbie, D. J. *J. Am. Chem. Soc.* **1987**, *109*, 2991.
20. Day, V. W.; Klemperer, W. G.; Yaghi, O. M. *J. Am. Chem. Soc.* **1989**, *111*, 4519.
21. Chen, Q.; Goshorn, D. P.; Scholes, C. P.; Tan, X.; Zubieta, J. *J. Am. Chem. Soc.* **1992**, *114*, 4667-4681.
22. Tourné, C.; Tourné, G.; Malik, S. A.; Weakley, T. J. R. *J Inorg. Nucl. Chem.* **1970**, *32*, 3875-3890.
23. (a) Mizuno, N.; Nozaki, C.; Kiyoto, I.; Misono, M. *J. Am. Chem. Soc.* **1998**, *120*, 9267-9272.; (b) *J. Catal.* **1999**, *182*, 285-288.
24. Pettersson, L.; Anderson, I.; Selling, A.; Grate, J. H. *Inorg. Chem.* **1994**, *33*, 982-993.

Chapter 13

Ultrastable Nanocapsules from Headgroup Polymerizable Divinylbenzamide Phosphoethanolamine

Glenn E. Lawson[1,*] and Alok Singh[2]

[1]Chemical and Biological Defense Center, Building 1480, Naval Surface Warfare Division, Dahlgren, VA 22448
[2]Center for Bio/Molecular Science and Engineering (Code 6900), Naval Research Laboratory, 4555 Overlook Avenue, S.W., Washington, DC 20375

Ultrastable nanocapsules have been produced by cross-linking a divinyl-benzamide phospholipid in the headgroup region. Based on ^{1}H NMR analysis, the cross-linking is complete in 5 min at room temperature. Freeze-drying the cross-linked vesicles produces a white powder which can be redispersed in water at room temperature. TEM analysis showed that the cross-linked vesicles were smooth, spherical, unilamellar, and of comparable size to the non-cross-linked vesicles. The robustness of the cross-linked vesicles was determined by subjecting the freeze-dried material to a series of treatments in aprotic, and protic organic solvents. Subsequent to treatment, TEM micrographs showed vesicles comparable in size and shape to the cross-linked vesicles. The chain melting behavior of the non-cross-linked phospholipids were 37.5 °C for non-cross-linked DPPE-DVBA, 29.4 °C for cross-linked DPPE-DVBA, 20.3 °C for non-cross-linked DLPE-DVBA and 13.9 °C for cross-linked DLPE-DVBA. The cooling phase transition for the non-cross-linked DPPE-DVBA was 32.8 °C and 25.9 °C for cross-linked DPPE-DVBA. Percent leakage of the water soluble dye carboxyfluorescein was determined for both the non-cross-linked and cross-linked DPPE-DVBA vesicles. The hydrophobic visible dye oil red-O was encapsulated in the vesicles after cross-linking and freeze-drying. Encapsulation was performed 20 °C above and below the phase transition of DPPE-DVBA phospholipid. The percentage of oil red-O encapsulated at 10 °C was 15% and the percentage oil red-O encapsulated at 55 °C was 17%, based on the starting weight of the oil red-O dye.

Introduction

Vesicles are unique self-assemblies that continually show great promise across numerous disciplines. Applications ranging from drug delivery (*1, 2*), to nanometer responsive materials (*3, 4*), and transportable catalytic nanocapsules have been explored (*5, 6*). In spite of this, the formation of highly stabilized vesicles has been elusive. This is, in part, due to the natural instabilities of conventional phospholipid vesicles. For instance, a serious weakness of phospholipid vesicles is that they are thermodynamically unstable and undergo secondary processes like fusion and disintegration. Upon exposure to organic solvents or detergents, their membrane integrity is destroyed (*7*). Attempts to stabilize the vesicular structure via the freeze-drying process result in fusion and extensive aggregation of vesicles (*8*). Overcoming these barriers could potentially extend the vesicle shelf life and increase their overall usage in technology development. Hence, strategies are continually being developed in an attempt to stabilize these physical and chemical instabilities (*9–11*). One method that has proven to be highly versatile is by stabilizing the vesicle structure through polymerization. This has been studied extensively in the hydrocarbon chain region and, to lesser extent, in the headgroup region (*12–15*). But, drawbacks are associated with polymerizing reactive groups in the bilayer membrane. For example, linear and cross-linking polymerizations are known to alter the natural properties of the membrane, resulting in the loss of the fluid phases. In comparison, linear polymerization in the headgroup region has afforded only limited stability (*16, 17*). Unlike non-polymerized vesicles, a recent study has shown that polymerized vesicles retain their original size following freeze drying (*18*). Thus, potentially extending the shelf life and increasing the possibility for new applications.

Non-phospholipid based materials are also known to form vesicles or capsules. These systems include polymersomes (*19, 20*), cereasomes (*21*), and alginate-based capsules (*22*). However, each of the systems has drawbacks, which range from particle size to accessibility of substrates. To remove these deficiencies, the approach of cross-linking the hydrophilic surface of small, unilamellar vesicles is attractive. It will increase the mechanical strength of the vesicle in the headgroup region, while retaining the natural properties of the membrane, thus, affording a stable hydrophilic interface that is permeable to a hydrophobic interior and a refillable central cavity.

In the present study, we report the efficient cross-linking of phospholipid vesicles in the headgroup region. The cross-linking has greatly enhanced their physical and chemical properties against the solubilizing power of organic solvents and the process of freeze-drying. The cross-linked phospholipid vesicles can entrap hydrophobic and hydrophilic components and remain technologically versatile for use across numerous disciplines. Figure 1 shows a schematic representation of a vesicle after cross-linking. The individual phospholipids consist of a divinylbenzoyl group conjugated in the headgroup region of 1,2-diacyl-*sn*-glycero-3-phosphoethanolamines (**1–3**); 1,2-dipalmitoyl-*sn*-glycero-3-phospho-*N*-(2-hydroxymethyl)-3,5-divinyl-benzamide (DPPE-DVBA, **1**), 1,2-dilauroyl-*sn*-glycero-3-phospho-*N*-(2-hydroxymethyl) 3,5-divinyl benzamide (DLPE-DVBA, **2**), and *N*-acyl-phosphotidylethanol-3,5-divinyl benzamide (EGGPE-DVBA, **3**).

Experimental

General

1,2-dipalmitoyl-*sn*-glycero-3-phosphoethanolamine, 1,2-dilauroyl-*sn*-glycero-3-phosphoethanolamine and L-α-phosphatidylethanolamine were purchased from Avanti Polar lipids. Unless stated otherwise, all solvents and reagents were purchased from Aldrich Chemical Company and used as received. Buffer was prepared by mixing a H_2O solution of 0.05 M $NaHCO_3$/0.1 M NaOH and adjusting the pH to 9.3. For thin layer chromatography (TLC), glass plates coated with silica gel-60, with fluorescent indicator, were used. For spectral characterization and analysis, Agilent 8453 UV-vis spectrophotometer, Bruker DRX-400 nuclear magnetic resonance spectrometer (400 MHz for proton and 100 MHz for carbon) and a Bruker Tensor 27 infrared spectrophotometer were used. Mass spectra of synthetic intermediates and lipids were recorded on an Applied Biosystems, MDS Sciex, API electrospray, QTOF, QSTAR® Pulsar instrument. Vesicle images were acquired using a Zeiss EM-10 transmission electron microscope equipped with a Spot Insight QE digital camera (Model 4.2) at 60 kV. Calorimetric studies on hydrated phospholipids and their mixtures were performed on DSC model 2920 Modulated DSC from TA Instruments. Fluorescence measurements were performed on a Jobin Yvon (Horiba) FluoroMax-3 instrument.

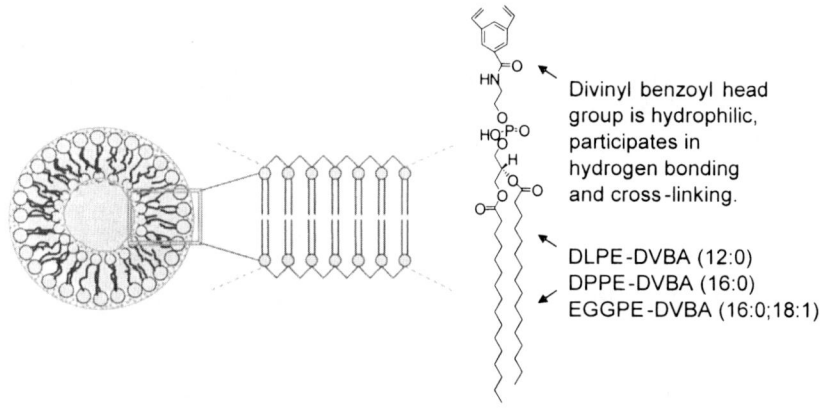

Figure 1. Schematic representation of cross-linked vesicle formed from DLPE-DVBA, DPPE-DVBA or EGGPE-DVBA. (See page 9 of color inserts.)

Synthesis of Phospholipids and their Intermediates

Synthesis of 3,5-divinylbenzoic acid (DVBA)

The general procedure has been previously described by Lawson (23). 3,5-divinylbenzoic acid; 310 mg (62%) yield. The acid was analyzed in a chloroform/methanol solution on silica gel coated TLC plates ($CHCl_3:CH_3OH$; 95:5 v/v). A single spot on TLC (R_f = 0.24) and a sharp melting point at 142–143 °C revealed the homogeneity of the product. ^1H NMR (400 MHz $CDCl_3$) δ 7.9 (s, 2H), 7.4 (s, 1H), 6.7 (dd, 2H, J = 11 Hz), 5.8 (d, 2H, J = 11 Hz), 5.3 (d, 2H, J = 18 Hz). ^{13}C NMR (100 MHz $CDCl_3$) δ 115.0, 126.0, 128.0, 130.0, 135.0, 137.0, 166.0. QTOF-MS calculated for $C_{11}H_{10}O_2$: 174.0681: Found 173.0660. FT-IR (KBr pellet) 3435, 2912, 2599, 1704, 1600, 1589, 1457, 1313, 1241 cm^{-1}.

Synthesis of 1,2-dipalmitoyl-sn-glycero-3-phospho-N-(2-hydroxyethyl)-3,5-divinylbenzamide (DPPE-DVBA) (1)

The general procedure has been previously described by Lawson (23). Coupling of DVBA with 1,2-dipalmitoyl-*sn*-glycerol-3-phosphoethanolamine; 88 mg (73%) yield. A single spot on TLC (R_f = 0.35) was detected using ($CHCl_3:CH_3OH$; 80:20, v/v) solvent system. Ninhydrin spray did not produce any stain on the plate indicating the absence of any free DPPE in the product collected after column chromatography. ^{1}H NMR (400 MHz CDCl$_3$) δ 7.90 (s (b), 2H), 7.40 (s (b), 1H), 6.61(d, 2H, J = 11 Hz), 5.82 (d, 2H, J = 11 Hz), 5.32 (d, 2H, J = 18 Hz), 4.10–3.51 (8H), 2.18 (s (b), 4H), 1.52–1.28 (s (b), 52 H), 0.92 (t, 6H). ^{13}C NMR (100 MHz CDCl$_3$) δ 17.0, 21.0, 22.0, 34.0–36.0, 37.0, 41.0, 61.0, 62.0, 118.0, 128.0, 130.0, 132.0, 138.0, 141.0, 168.0, 172.0. QTOF-MS calculated for C$_{48}$H$_{82}$NO$_9$P: 847.5727: Found 848.5797. FT-IR (KBr pellet): 3380, 2942, 2858, 1746, 1656, 1595, 1547, 1247, 1079 cm^{-1}.

Synthesis of 1,2-dilauroyl-sn-glycero-3-phospho-N-(2-hydroxyethyl)-3,5-divinylbenzamide (DLPE-DVBA) (2)

The general procedure has been previously described by Lawson (23). Coupling of DVBA with 1,2-dipalmitoyl-*sn*-glycerol-3-phosphoethanolamine; 20 mg (63%) yield. A single spot on TLC (R$_f$ = 0.64), silica gel, (CHCl$_3$:CH$_3$OH; 80:20, v/v). Ninhydrin spray did not produce any stain on the plate, indicating the absence of any free DLPE in the product collected after column chromatography. ^{1}H NMR (400 MHz CDCl$_3$) δ 7.91 (d, 2H), 7.54 (d, 2H), 6.63 (dd, 2H, J = 11.0 Hz), 5.81 (d, 2H, J = 18.0 Hz), 5.3 (d, 2H, J = 11.0), 5.22 (s (b), 1H), 4.33, 4.12, 3.97, 3.79 (m, 8H), 2.18 (s (b), 4H), 1.52-1.28 (s (b), 36H), 0.92 (t, 6H). ^{13}C NMR (100 MHz CDCl$_3$) δ 17.0, 21.0, 22.0, 34.0–36.0, 37.0, 41.0, 61.0, 62.0, 118.0, 128.0, 130.0, 132.0, 138.0, 168.0, 172.0. QTOF-MS calculated for C$_{40}$H$_{65}$NO$_9$P: 734.4397: Found 735.3078. FT-IR (KBr pellet): 3374, 2930, 2858, 1740, 1638, 1583, 1541, 1475, 1240, 1081 cm^{-1}.

Synthesis of N-acyl-phosphotidylethanol-3,5-divinylbenzamide (EGGPE-DVBA) (3)

The general procedure has been previously described by Lawson (23). Coupling of DVBA with L-α-phosphatidylethanolamine; 88 mg (73%). A single spot on TLC (R$_f$ = 0.32, silica gel, (CHCl$_3$:CH$_3$OH; 80:20, v/v). Ninhydrin spray did not produce any stain on the plate, indicating the absence of any free EGG-PE in the product collected after column chromatography. ^{1}H NMR (400 MHz CDCl$_3$) δ 7.90 (s (b), 2H), 7.40 (s (b), 1H), 6.61 (d, 2H), 5.82 (d, 2H), 5.32 (d, 2H), 5.23 (s (b) 1H), 5.22 (s (b) 2H), 4.10–3.51 (8H), 2.77 (s (b) 4H), 2.18 (s (b), 4H), 1.52–1.28 (s (b), 46 H), 0.92 (t, 6H). ^{13}C NMR (100 MHz CDCl$_3$) δ 17.0, 21.0, 22.5, 23.0, 34.0–36.0, 37.0, 41.0, 61.0, 62.0, 70.5, 118.0, 128.0, 130.0, 132.0, 138.0, 141.0, 168.0, 172.0. QTOF-MS calculated for C$_{50}$H$_{84}$NO$_9$P: 873.5884: Found 874.4236. FT-IR (KBr pellet): 3332, 2930, 2852, 1740, 1644, 1602, 1553, 1463, 1241, 1073 cm^{-1}.

Characterization of Phospholipid Physical Properties

Differential Scanning Calorimetric (DSC) Studies.

Phospholipid samples were weighed directly in the DSC pans for simplicity and convenience. The phospholipid was dissolved in 500 μL chloroform, resulting in a homogenous solution. The solvent was then removed under a gentle stream of nitrogen so that a uniform thin film would be formed, followed by applying a high vacuum for at least 3 h to remove traces of solvent. A thin lipid film was covered with 15 μL deionized water and the pans were immediately sealed and left in an oven set at 70 °C for 5 h to allow complete hydration. For cross-linked samples, freeze-dried, cross-linked vesicles were transferred into DSC pans and weighed. The freeze-dried, cross-linked material was hydrated with 15 μL deionized water. The pans were then immediately sealed and left in an oven at 70 °C for 5 h to allow complete hydration before running DSC scans.

Preparation of Langmuir Film

The lipids were dissolved separately in chloroform. Monolayers of the pure lipids were spread as a monolayer film on a Langmuir-Blodgett mini trough (total area of the trough is 85 cm^2) from NIMA Technology (model 601B), using a 10 μL Hamilton syringe, with water as the subphase. Isotherms were generated using a computerized data acquisition system. The film pressure was measured using a Wilhelmy balance with platinum foil as the plate. Monolayers were compressed at a rate of 10 cm^2/min to their collapse pressures and were reproducible. The pressure-area isotherms were reproducible in successive runs.

Preparation and Characterization of Vesicles

Preparation and Visualization of Vesicles

The phospholipid concentration in a chloroform solution was 10 mg/mL. A thin lipid film was coated on the walls of the tube by removing solvent under a gentle stream of nitrogen, followed by thorough drying of the film under high vacuum. Thin lipid films were covered with 0.05 M $NaHCO_3$/0.1 M NaOH buffer at a pH value of 9.3 so that vesicle formation is optimized. Lipid dispersions were hydrated by heating at 50 °C for 1 h and dispersed in the medium by intermittent vortex mixing followed by sonication at 50 °C using a

Branson Sonifier-model 450. A cup horn device equipped with water intake and outlet connections was used for sonicating the sample. In all of the cases, 50% power and 80% duty cycle, with 1 h sonication, produced dispersions of constant turbidity at 400 nm (UV-vis, ABS 0.02), leading to the formation of uniform sized vesicles.

For TEM analysis, a drop of vesicle suspension was placed on 200-mesh copper formvar/carbon grid. Vesicles on the grid were stained by placing a drop of 1% aqueous uranyl acetate solution followed by removal of excess solution by wicking it with a piece of filter paper. The vesicle size was determined by counting and measuring 100 vesicles at random.

Cross-linking of Vesicles

All cross-linking experiments were carried out at room temperature, as follows. Vesicles were cross-linked using a combination of water-soluble radical initiation and photo cross-linking. Cross-linking was monitored until complete, using TLC and NMR techniques. For all NMR studies, the vesicles were prepared in 0.7 mL of D_2O solution of 0.05 M $NaHCO_3$/0.1 M NaOH and adjusting the pH to 9.3. For radical initiated cross-linking, the water-soluble initiator, 2,2'-azobis(2-methylpropionamidine) dihydrochloride (AAPD), was used in combination with UV irradiation at 254 nm. Vesicle dispersions and AAPD solutions were thoroughly degassed with nitrogen for 25 min to exclude oxygen. AAPD was added to obtain the desired mole ratio of 7:1, 25:1, or 100:1 (phospholipids:initiator) molar ratio. Photochemical reactions were carried out in a quartz cell placed 12 cm distance from the light source, using a Rayonet photochemical reactor. The cross-linking progress was monitored by recording ^1H NMR spectra at different time intervals. TLC characterization was performed using the solvent system (chloroform:methanol:water, 65:25:4). The $R_f = 0.0$ indicated complete cross-linking.

Stability of Cross-linked Vesicles

Dispersions of cross-linked vesicles were prepared from above procedures with DPPE-DVBA or EGGPE-DVBA in a 30 mg/mL concentration. The dispersions were quickly frozen by immersing in a dry ice/ethanol bath and then subjected to high vacuum, overnight, using a Labconco freeze-dry system (model Freezone 4.5). Cross-linked, freeze-dried DPPE-DVBA or EGGPE-DVBA vesicles were divided into two equal parts. One part was redispersed in deionized water at room temperature with intermittent vortex mixing and then examined by transmission electron microscopy. The second part of the freeze-dried, cross-linked vesicles were treated individually (~2 mg cross-linked powder/solvent), for 2 h, with each of the organic solvents (acetone, chloroform, ethanol (95%), hexane, methanol, and methylene chloride), followed by slow evaporation under a gentle stream of nitrogen and drying for at least 3 h under high vacuum. The vesicles were then redispersed in water before examination under a transmission electron microscope.

Carboxyfluorescein Entrapment and Leakage from DPPE-DVBA Vesicles

Non-cross-linked vesicles and cross-linked vesicles were prepared as stated under the section on preparation of DPPE-DVBA vesicles. Carboxyfluorescein (CF), 50 mM in 10 mM Tris buffer pH 8.0, was added to freeze-dried DPPE-DVBA phospholipid to make 5 mg/mL suspension. Sonication at 50% power and 80% duty cycle with 1 h sonication at 50 °C produced dispersions of constant turbidity at 400 nm (UV-vis, ABS 0.02), leading to the formation of uniform sized vesicles. Exogenous CF was separated from the DPPE-DVBA-CF entrapped vesicles on a Sephadex column (gel size 50–150) and eluted with 10 mM Tris buffer of pH 8.0. Fluorescence-release measurements were taken at 22 °C. Leakage of non-cross-linked and cross-linked DPPE-DVBA vesicles was determined with an excitation and emission wavelength of 492 and 520 nm. CF release was activated by addition of 10% TX-100 (Triton X-100), which solubilizes the vesicle membranes, hence releasing the entrapped CF contents. At the moment CF is released from the vesicles, the fluorescence intensity increases due to the dilution of CF (CF self-quenches at high concentrations). The leakage of CF from the DPPE-DVBA was calculated, where F_t is the fluorescence at time (t), F_o is the fluorescence at time (0), and F_{TX-100} is the fluorescence after Triton X-100 addition.

$$\% \ leakage = \frac{F_t - F_0}{F_{TX-100} - F_t} \times 100 \tag{1}$$

Hydrophobic Dye Encapsulation in Cross-linked Vesicles

In the hydrophobic encapsulation experiments, phospholipid DPPE-DVBA was used as representative lipid. The encapsulation experiment was conducted at two separate temperatures, 55 °C and at 10 °C [DPPE-DVBA phospholipid (2.5 mg/temperature), 5 mg total]. DPPE-DVBA was dissolved in chloroform to form a homogenous solution, then the solvent was evaporated under a gentle stream of nitrogen. The thin layer was dried under vacuum for at least 3 h. The thin lipid film was covered with 0.05 M $NaHCO_3$/0.1 M NaOH. Lipid dispersions were hydrated by heating at 50 °C for 1 h and dispersed in the medium by intermittent vortex mixing followed by sonication at 50 °C using a Branson Sonifier-model 450. Sonication for 1 h produced dispersions of constant turbidity, as observed using UV-vis, with a recurring absorbance at 400 nm (UV-vis, ABS 0.02). The vesicle suspension of uniformed-sized vesicles was characterized by TEM, as described under visualization of vesicles. Cross-linking was performed using the 25:1 molar ratio of (phospholipids:initiator) and using the protocol under cross-linking of vesicles. Characterization for polymerized vesicles was conducted by TLC. The sample was then freeze-dried overnight, leaving a white powder. Freeze-dried, cross-linked vesicles were divided into two separate samples (2.5 mg) each and placed into separate vials. The hydrophobic dye oil red-O (0.9 mg, 1 mole equivalent to 2.5 mg of lipid) was dissolved in 1 mL ethanol/water (1:1 v/v). The dye (0.5 mL) was added to each of the sample vials of freeze-dried, cross-linked vesicles. One was incubated overnight at 55 °C and the other at 10 °C overnight. The suspension of cross-linked vesicles containing encapsulated oil red-O dye were separated from free dye by isolating a cross-linked vesicle pellet after centrifuging at 5000 g for 30 min. The clean solution was removed and the pellet washed with ethanol/water (1:1 v/v). The clean solution was tested for cross-linked vesicles by TLC analysis with the solvent system (chloroform:methanol:water 65:25:4). Concentration of encapsulated oil red-O was determined through UV-vis analysis. The oil red-O encapsulated, cross-linked pellet was suspended in 3 mL ethanol/water mixture (1:1 v/v) followed by dispersing with bench top sonication for 5 min to release the encapsulated dye. UV-vis analysis was performed and the absorbance recorded. A series of known concentrations of oil red-O dye in an ethanol/water mixture (1:1 v/v) were used to determine the oil red-O encapsulated concentration. The data were then plotted as absorbance versus concentration.

Results and Discussion

The focus of this study was to explore the efficiency and extent of cross-linking that can be obtained under mild conditions. It is known that polymerization can cause phase separation of the membrane, resulting in destabilization or fusion of the vesicular structure. Therefore, we wanted to maximize cross-linking in the headgroup region, while retaining the membrane properties. Thus, stable vesicles are produced that demonstrate enhanced strength against mechanical, physical or chemical perturbants. Under mild experimental conditions, complete cross-linking is obtained, as shown in Figure 2A. We believe that this afforded long, cross-linked chains on the phospholipid hydrophilic surfaces of the inner aqueous compartment and outer periphery of the vesicle. It is also reasonable to assume that incomplete cross-linking would lead to short intermittent chains of various lengths (Figure 2B). This would be verified easily by the incomplete reduction of the vinylic proton signals shown in ^{1}H NMR analysis. It is known from transmission electron micrographs (TEM) taken after cross-linking that disruption of the vesicle is not occurring, as depicted in Figure 2C. This is strong experimental evidence that the cross-linking performed in this study is a non-disruptive process to the structural integrity of the vesicles. The vesicle sizes before and after cross-linking are similar, demonstrating the reproducibility of our general protocol. The robustness of the cross-linked vesicles is demonstrated by retention of the vesicular structure after freeze-drying and redispersion in water. This enhances the utility of our protocol by increasing the shelf life in a powder form. TEM results after redispersion of the freeze-dried, cross-linked vesicles show that the vesicles are resistant to the solubilizing action of organic solvents. In addition, we are also optimizing methods for entrapment of either non-polar or aqueous soluble encapsulants in the cross-linked vesicles.

Surface Pressure-Area (π-A) Isotherms

The surface pressure-area isotherms of phospholipids 1 and 2 on water are shown in Figure 3. The referenced isotherms are dipalmitoylphosphatidylethanolamine (DPPE) (24), dipalmitoylphosphatidylcholine (DPPC) (25), dilauroylphosphatidylethanolamine (DLPE) (26), and dilauroylphosphatidylcholine (DLPC) (27). Isotherm of phospholipid 1 has a limiting molecule area of 80 Å^2/molecule and a collapse pressure of 69 mN/m, as compared to 70 Å^2/molecule and 45 mN/m, respectively, for phospholipid 2. The monolayer of phospholipid 1 lifted off with a larger area than did the monolayer of phospholipid 2. Monolayers of phospholipids 1 and 2 are both compressible at 60 Å^2/molecule. It can be seen that no liquid expansion/liquid compression (LE/LC) transition is present for either phospholipid 1 or 2. This is the onset of a phase transition from a fluid phase to a more highly ordered condensed phase, which is typically not seen in phospholipid isotherms. Since the headgroups of phospholipids 1 and 2 are identical, the differences in the monolayer properties of these lipids at the collapsing pressure should be due to the difference in the number of carbons in the acyl chains. We believe this can be attributed to the extra energy needed to align the chains in the dipalmitoylphospholipids.

In addition, the packing of the headgroup also affects the packing differences of the phospholipids. The monolayer of phospholipid 1 collapses 16 °C higher than DPPE or DPPC. On the other hand, the collapse pressure of phospholipid 2 is similar to the reference phospholipids DLPE and DLPC. Furthermore, the thermal behavior of phospholipids 1 or 2 suggest that well-organized bilayers are not being formed where the headgroups are tightly packed and the acyl chains are well-aligned.

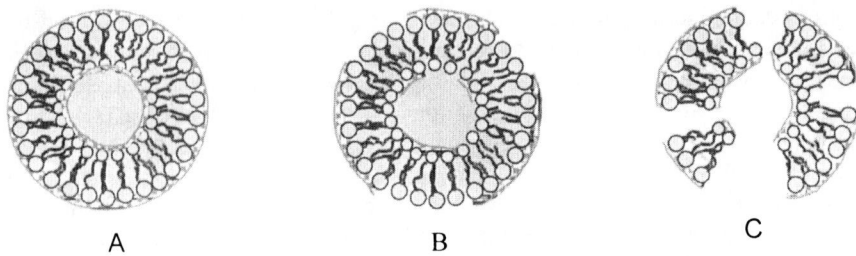

Figure 2. Graphic representation of vesicle cross-linking: (A) complete cross-linking; (B) partial cross-linking; (C) destructive cross-linking to vesicle structure. (See page 9 of color inserts.)

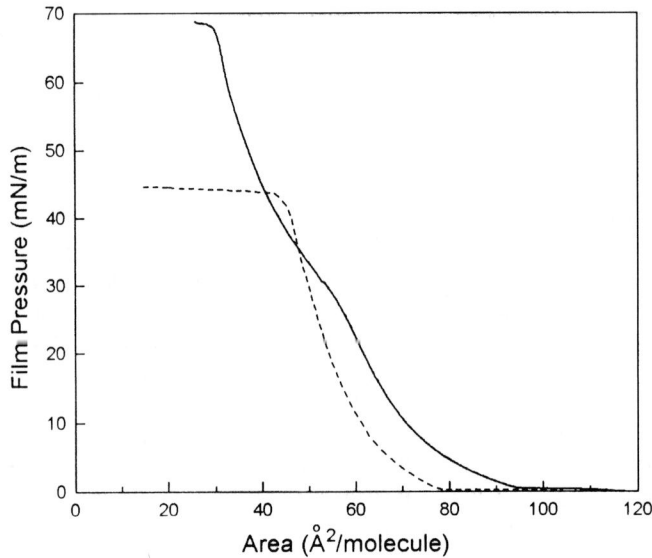

Figure 3. Surface pressure-area (π-A) isotherms of DPPE-DVBA (—) and DLPE-DVBA (----) at 20 °C.

Vesicle Dispersion

Formation of stable, unilamellar vesicles for phospholipids **1–3** was produced with mild sonication at 50 °C for 1 h and was optimized for unilamellar vesicle formation at a pH of 9.3. A clean, almost translucent dispersion was easily obtained under the previously described conditions (*23*). Monitoring the dispersion using UV-vis at 400 nm for turbidity resulted in an absorbance lower limit of 0.02. Freeze-drying the cross-linked vesicles produced a white powder. Redispersion of the cross-linked, freeze-dried powder in either water or buffer resulted in a clear dispersion by simple vortex mixing or bench top sonication for 5–10 min.

Cross-linked Vesicles

Control studies were conducted with either phospholipid **1** alone or phospholipid **1** with 2,2′-azobis(2-methylpropionamidine)dihydrochloride (AAPD) in the molar ratio of phospholipids:initiator at 7:1. To ensure that the membrane remained in the fluid phase, the temperature was maintained at 50 °C. All solutions were thoroughly degassed with nitrogen to ensure there were no contributions from oxygen radicals. The thermally-induced, cross-linking of phospholipid **1** with AAPD alone at 50 °C was determined by proton NMR integration. The extent of cross-linking was shown to be only a 50% conversion in 1.5 h. A control experiment for UV alone induced cross-linking was conducted and TLC was used to test the progress of the reaction. At the end of five hours, TLC still showed the presence of starting phospholipid. In addition, a slower moving fraction was assigned to oligomers, and a non-mobile spot located at the point of origin was assigned to polymeric vesicles. ^{1}H NMR confirmed through integration that the cross-linking was only 48% complete. The efficiency of the cross-linking was studied by varying the amount of the water-soluble radical initiator, AAPD while maintaining a constant amount of the phospholipids. Cross-linking in vesicles was performed in D_2O. Before addition of an aqueous solution of AAPD, the vesicle dispersion was thoroughly degassed of oxygen by bubbling nitrogen. The initial concentration of (phospholipids:initiator) was a 7:1 molar ratio. The cross-linking of the vesicles composed of phospholipid **1** in the UV irradiation was finished after five min at room temperature. ^{1}H NMR analysis conducted immediately following irradiation showed the complete disappearance in the intensity of vinyl proton signals at 6.7, 5.8, and 5.3 ppm.

Efficiency of the cross-linking with phospholipid **1** was tested by increasing the molar ratio of (phospholipids:initiator) from 7:1 to 25:1. Again, the vesicle dispersion and AAPD solution were degassed thoroughly with nitrogen before mixing. Figure 4A shows that vesicles produced from phospholipid **1**, with a 16-carbon long dipalmitoyl chains, were completely cross-linked at room temperature in 5 min of UV irradiation at 254 nm. In comparison, vesicles produced from phospholipid **3** (not shown), which contain a palmitoyl chain and an unsaturated oleoyl chain, were also completely cross-linked at room temperature in 5 min. ^{1}H NMR analysis was conducted immediately following the irradiation and showed the complete disappearance in the intensity of vinyl proton signals at 6.7, 5.8, and 5.3 ppm. Vesicles formed from phospholipid **2**, which contain two 12-carbon long dilauroyl chains, were completely cross-linked in 5 min, as shown in Figure 4B. ^{1}H NMR analysis performed immediately after the UV irradiation showed the complete reduction in the intensity of the vinyl proton signals at 6.7, 5.7, and 5.2 ppm

In an attempt to optimize the lower limit concentration of AAPD, vesicles formed from phospholipid **1**, alone, were tested for the extent of cross-linking by increasing the molar ratio of (phospholipids:initiator) to 100:1. Vesicles produced from phospholipid **1** showed that, after 5 min, only 79% of the phospholipid in the vesicles was cross-linked, as determined ^{1}H NMR analysis. An additional 10 min of UV exposure increased the cross-linking extent to 92%. Thereafter, it took an additional 10 min of UV irradiation for complete disappearance of the proton signals at 6.6, 5.8, and 5.3 ppm, indicating complete cross-linking of the individual phospholipids in the vesicles. Therefore, the 25:1 molar ratio of (phospholipids:initiator) was taken as the lower concentration limit of AAPD in order to obtain 100% cross-linking in 5 min at ambient temperature for phospholipids **1–3**.

Thermotropic Behavior of Hydrated Phospholipids

Differential scanning calorimetry was performed to measure the chain melting temperature for non-cross-linked phospholipids (NCP) **1–3**. Non-cross-linked vesicle dispersions (NCVD) and cross-linked vesicle dispersions (CVD) of **1** and **2** were also measured along with the chain melting transition of CVD-**1** with a hydrophobic dye entrapped. Parent phospholipids DPPE and DLPE were used as references for NCP **1** and **2**. Parent phospholipid EGGPE was used as reference for phospholipid **3**. NCPs were weighed directly into the DSC pans. Vesicle dispersions were used for the non-cross-linked vesicle samples and freeze-dried powder was used for the cross-linked vesicle dispersion samples.

Table I shows the thermotropic transitions of phospholipids and vesicle dispersions of **1** and **2**, along with enthalpies of transition. NCP-1 shows a sharp chain melting transition temperature at 37.5 °C and a sharper transition peak at 32.8 °C upon cooling. The ΔH at 37.5 °C was determined to be 6.07 kcal/mole. Dispersions of the NCVD-1 show a chain melting transition at 25.9 °C and a cooling transition at 23.6 °C. CVD-1 shows a melting transition at 29.4 °C and a cooling transition at 25.9 °C.

The ΔH was shown to be 2.50 kcal/mole at 29.4 °C. The NCP-1 chain melting and cooling transitions are within 12 °C of the NCVD-1 and CVD-1 transitions. In addition, both melting and cooling transition temperatures of NCVD-1 and CVD-1 are within 4 °C of each other. The chain melting transition of parent DPPE phospholipid is at 63 °C, which is also comparable, indicating that the acyl chain membrane properties are not seriously affected by the cross-linking. NCP-2, Table I, shows a sharp chain melting transition at 20.3 °C, but no cooling transition could be seen. The ΔH at 20.3 °C was shown to be 5.52 kcal/mole. In Table I, NCVD-2 (not shown) shows a melting transition at

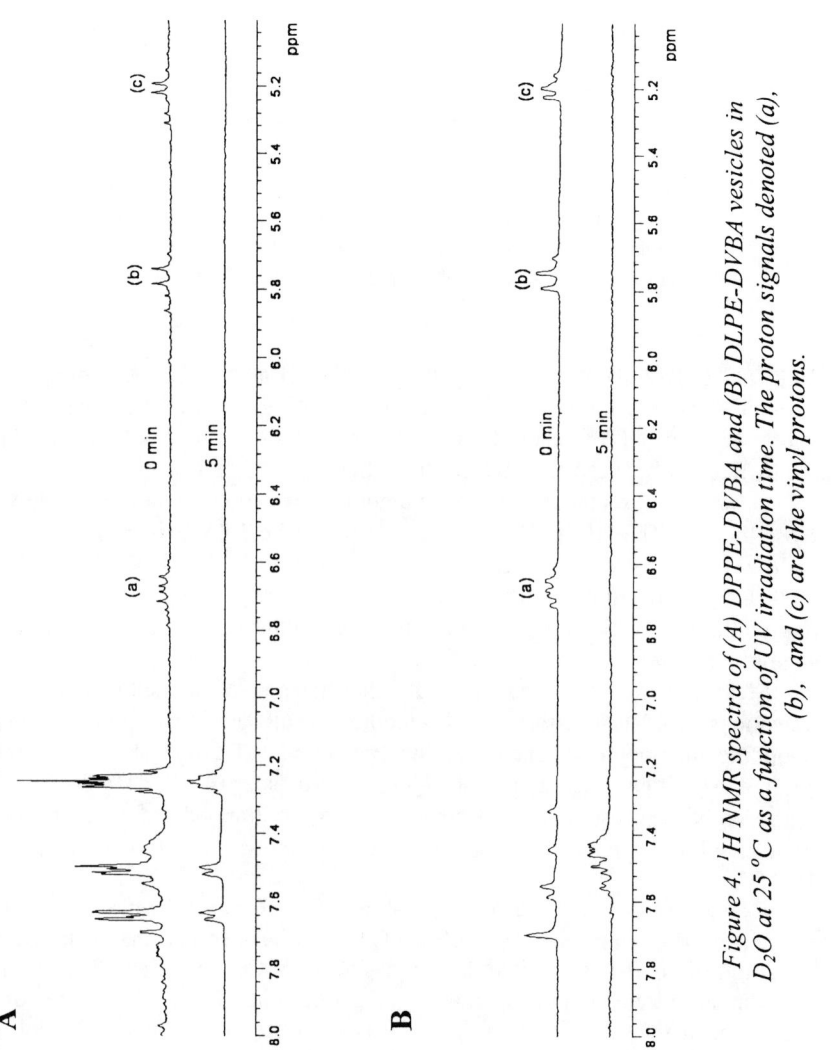

Figure 4. ^1H NMR spectra of (A) DPPE-DVBA and (B) DLPE-DVBA vesicles in D_2O at 25 °C as a function of UV irradiation time. The proton signals denoted (a), (b), and (c) are the vinyl protons.

Table I. Transition Properties of DPPE-DVBA and DLPE-DVBA Phospholipids and Vesicles

Phospholipid	T_m (°C)	T_c (°C)	ΔH (kcal/mole)
NCP[a]-1	37.5	32.8	6.07
NCVD[b]-1	25.9	23.6	----
CVD[c]-1	29.4	25.9	2.50
CVD[c]-1-oil-red-O	15.3	13.1	0.055
NCP[a]-2	20.3	----	5.52
NCVD[b]-2	10.8	----	----
CVD[c]-2	13.9	----	1.05

[a] NCP is an abbreviation for non-cross-linked phospholipids, [b] NCVD is an abbreviation for vesicle dispersion non-cross-linked, [c] CVD is an abbreviation for cross-linked vesicle dispersion.

10.8 °C. CVD-2 shows a sharp melting transition at 13.9 °C. A cooling phase transition was not found. The NCP-2 chain melting and cooling transitions are within 10 °C of NCVD-2 and CVD-2 melting transitions. It should be noted that the difference between the melting transitions of NCVD-2 and CVD-2 is only 3.1 °C, which shows that the bilayer membrane properties are not affected by the cross-linking. The ΔH at 13.9 °C was shown to be 1.05 kcal/mole. The chain melting phase transition of the parent phospholipid DLPE is also comparable at 29.0 °C. As a result of the natural properties of phospholipid 3, neither heating nor cooling transitions were found for the non-cross-linked or cross-linked samples.

Measurements were recorded on the thermotropic chain melting and cooling transitions of CVD-1 with the hydrophobic dye oil red-O entrapped within the lipophilic membrane. A chain melting transition was found at 15.3 °C and a cooling transition was seen at 13.1 °C. There is a 14 °C difference in the transition of the cross-linked vesicles without dye entrapped and the cross-linked vesicles with dye entrapped. The ΔH value was shown to be 0.055 kcal/mole at 15.3 °C.

Examining the DSC results in Table I for bilayer formation, the DPPE reference phospholipid has a T_m of 63 °C. It can be seen that the chain melting transition of NCP-1 is lowered 25 degrees to 37.5 °C. We believe this lowering in the melting transition of the bilayer is a result of an increase in the area of the headgroup from an amine to a divinylbenzoyl group. In relation to vesicles, it has been shown that polymerization of the hydrocarbon chains usually leads to loss of a well-defined chain melting transition (28). In contrast, polymerization in the hydrophilic head group region generally does not significantly affect the melting transitions of the hydrocarbon chain (17). It is also reasonable to assume that, after polymerization in the headgroup region, the packing of the planar aromatic headgroup would increase. This increase in headgroup packing

(NCVD-1 to CVD-1) can be seen in the chain melting temperatures in Table I. After formation of the NCVD-1, a lowering of the T_m can be noticed from 37.5 to 25.9 °C, a decrease of 12 degrees. This can be explained by formation of small, unilamellar vesicles which experience a significant amount of curvature strain due to their small size. This was reversed after cross-linking occurred forming CVD-1. The T_m of NCVD-1 increases from 25.9 °C to 29.4 °C. This is an indication that cross-linking in the head group region has resulted in tighter packing. Moreover, this asserts a small organizational influence on the acyl chain packing in the bilayer, thus, an increase in chain melting transition. In view of the above argument, the T_m should be lowered after entrapment of the hydrophobic dye oil red-O in the membrane of CVD-1. It can be seen that CVD-1 before entrapment shows a T_m of 29.4 °C. Following entrapment, the T_m is 15.3 °C. We believe this is a result of the acyl chain organization being disrupted because of the bulk portion of the dye. On the molecular scale, it is reasonable to assume that the bulkiness of the aromatic groups initiated the disorganization because of free rotational movement of the molecule in a fluid phospholipid membrane.

A similar set of observations can be seen in the phospholipid **2** case. The DLPE reference phospholipid has a T_m of 29 °C. A lowering of 9 degrees is observed for the chain melting transition from 29 °C to 20.3 °C for NCP-**2**. After forming unilamellar vesicles of NCVD-**2**, the chain melting transition is decreased by 10 degrees from 20.3 °C to 10.8 °C. An increase by 3 °C is seen after formation of CVD-**2** through cross-linking in the headgroup region.

Transmission Electron Microscopic Visualization

Vesicles were observed by negative staining TEM. The non-cross-linked and cross-linked samples for phospholipids **1** and **3** are shown in Figures 5(A-B) and 6(A-B) for verification and also for comparison of the vesicle structure. Examination of the micrographs indicate that the vesicles are smooth, spherical, unilamellar and abundant in large vesicle populations. After cross-linking, TEM micrographs, show that the appearance and average size of the vesicles are not seriously altered. For phospholipid **1**, the non-cross-linked and the cross-linked vesicles have an average size of 70 nm. The non-cross-linked and cross-linked vesicles from phospholipid **3** show an average size of 70 nm.

Stability of Cross-linked Vesicles

The mechanical and chemical stability of the cross-linked vesicles were analyzed by negative staining TEM after exposure to protic and aprotic organic solvents. The solubilizing effects of organic solvents are known to destroy the structural integrity of conventional vesicles. We have shown that the cross-linked vesicles of phospholipid 1 could be prepared in water, freeze-dried to produce a white powder, and then redispersed back into water with no apparent change in the size or appearance of the nanostructure of the vesicle. In the present study, we have examined the stability of the cross-linked vesicles by treating the freeze-dried white powder with a series of common organic solvents, followed by redispersion in water after evaporating the organic solvent with a gentle stream of nitrogen and drying under high vacuum for at least three hours. Some vesicle dispersions showed aggregation at first, but a clean, almost translucent, homogenous suspension could be obtained with the use of agitation for five minutes from a bench-top sonicator. This is similar to the experimental protocol for reduction of aggregates reported by Liu and O'Brien (18). Negative staining TEMs of the cross-linked vesicles from phospholipid 1 showed the vesicles robustness by retaining their unilamellar, smooth and spherical appearance. It should be noted, as shown in Table II, that the greatest size differences between the vesicles before and after treatment with organic solvents are only around ten nanometers. Figure 5C shows images of the vesicles after being treated with the solvent acetone. A high population of vesicles can be seen in the micrograph at a size of 58 nm. Treatment with chloroform, a phospholipid solubilizing solvent (Figure 5D), the TEM images of the vesicles are 57 nm in size.

Images in Figure 5E were captured after treatment with the solvent ethyl alcohol (95%) which shows images of the vesicles at 66 nm in size. Following treatment with a lipophilic solvent hexane (Figure 5F), images of the vesicles are 62 nm in size. In Figure 5G, treatment with the solvent methanol shows an average size of vesicles images at 66 nm. After treatment with methylene chloride in Figure 5H, the size of vesicle images were 66 nm.

Negative staining TEMs of the cross-linked vesicles from phospholipid 3 also showed that the vesicles retained a unilamellar, smooth and spherical structure. Examination of the cross-linked EGGPE-DVBA vesicles also showed enhanced stability against treatment with organic solvents. In Figure 6C, images of the vesicles after treatment with acetone show vesicles at a size of 80 nm. Following treatment with chloroform (Figure 6D), the size of vesicle images were 62 nm. Figure 6E shows images of the vesicles with an average size of 71 nm after treatment of with ethyl alcohol 95%.

Subsequent to treatment with hexane, as shown in Figure 6F, the vesicles have an average size of 67 nm. In Figure 6G is shown images following treatment with methanol. Images of the vesicles were 62 nm. Following treatment with methylene chloride, the size of the vesicles was 66 nm. The solubilizing power of organic solvents against non-polymerized and polymerized vesicles is known to represent extremely harsh conditions (8). The results here demonstrate the practicability of recovering cross-linked vesicles from the process of freeze-drying and exposure to common organic solvents. These results also reveal that vesicles with cross-linking in the headgroup can be flexible and still remain highly stable three-dimensional objects or "nanocapsules".

Table II. Diameter of Cross-linked Vesicles after Treatment with Solvents

Solvents	Diameter (nm), 1	Diameter (nm), 2
Water, before cross-linking	70	70
Water, after cross-linking	70	70
Acetone	58	80
Chloroform	57	62
Ethyl alcohol	66	71
Hexane	62	67
Methanol	66	62
Methylene Chloride	66	66

Leakage of Carboxyfluorescein from DPPE-DVBA Vesicles

The leakage percent versus time shown in Figure 7 for the cross-linked DPPE-DVBA and non-cross-linked DPPE-DVBA vesicles are comparable. Carboxyfluorescein (CF) leakage after 15 min for the non-cross-linked system was 60%, which increased to 79% CF leakage at 30 min and complete leakage of all CF at 60 min. The DPPE-DVBA cross-linked vesicles were similar with 50% CF leakage at 15 min, then increasing to 84% at 45 min and 100% at 60 min.

Entrapment of Oil Red-O in Freeze-dried Cross-linked Vesicles

Encapsulation of hydrophobic dye oil red-O was performed 20 °C below the chain melting transition of phospholipid 1 (37.5 °C) and 25 °C above the chain melting transition. The cross-linked vesicles with encapsulated oil red-O were separated from the free non-encapsulated oil red-O by centrifuging, followed by washing. After isolation of the oil red-O encapsulated cross-linked vesicle pellet, UV-vis analysis was performed in order to determine the percentage of encapsulation of the oil red-O in the cross-linked vesicles. The percentage of oil red-O encapsulated at 10 °C was 15% and the percentage oil red-O encapsulated at 55 °C was 17%, based on the starting weight of the oil red-O. This data suggests that it is possible to recycle stabilized vesicles by filling and refilling as required at variable temperatures.

Conclusions

In this paper, we have emphasized that complete cross-linking in the headgroup region is efficiently achieved with AAPD at ambient temperature

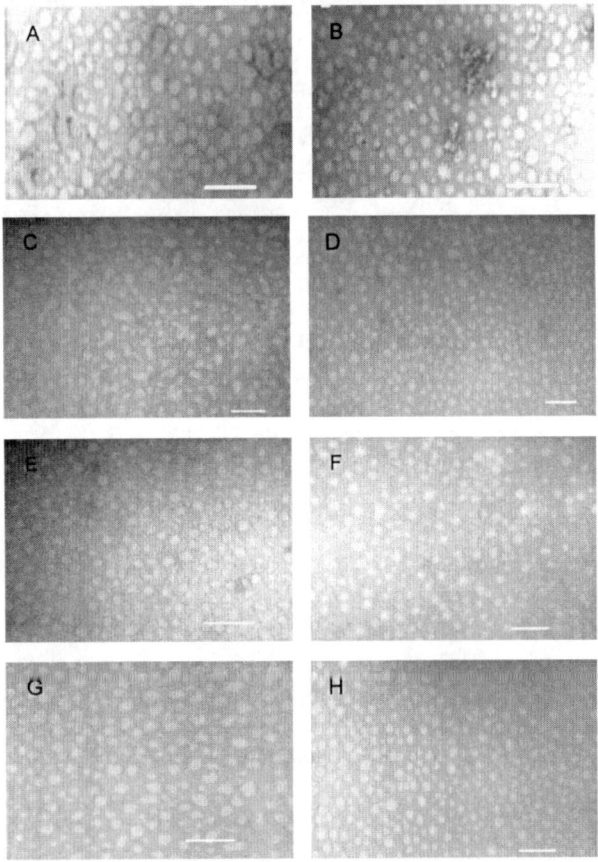

Figure 5. Transmission electron micrographs of vesicles prepared from DPPE-DVBA. The scale bar is 250 nm in length. (A) DPPE-DVBA before cross-linking; (B) DPPE-DVBA after cross-linking; (C) DPPE-DVBA after cross-linking, followed by freeze-drying and treatment with acetone; (D) DPPE-DVBA after cross-linking, followed by freeze-drying and treatment with chloroform; (E) DPPE-DVBA after cross-linking, followed by treatment with ethanol; (F) DPPE-DVBA after cross-linking, followed by freeze-drying and treatment with hexane; (G)DPPE-DVBA after cross-linking, followed by freeze-drying and treatment with methanol; (H) DPPE-DVBA after cross-linking, followed by freeze-drying and treatment with methylene chloride.

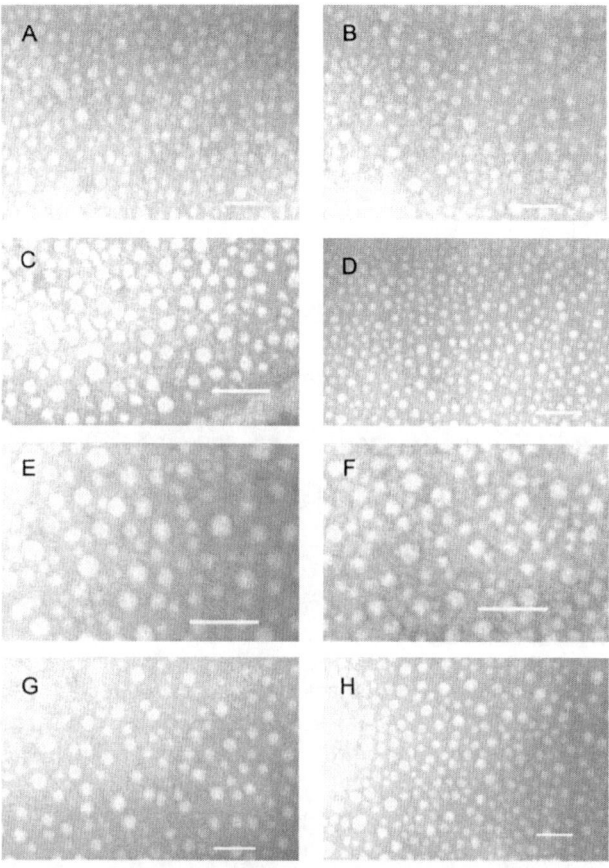

Figure 6. Transmission electron micrographs of vesicles prepared from EGGPE-DVBA. The scale bar is 250 nm in length. (A) EGGPE-DVBA before cross-linking; (B) EGGPE-DVBA after cross-linking; (C) EGGPE-DVBA after cross-linking, followed by freeze-drying and treatment with acetone; (D) EGGPE-DVBA after cross-linking, followed by freeze-drying and treatment with chloroform; (E) EGGPE-DVBA after cross-linking, followed by freeze-drying and treatment with ethanol; (F) EGGPE-DVBA after cross-linking, followed by freeze-drying and treatment with hexane; (G) EGGPE-DVBA after cross-linking, followed by freeze-drying and treatment with methanol; (H) EGGPE-DVBA after cross-linking, followed by freeze-drying and treatment with methylene chloride.

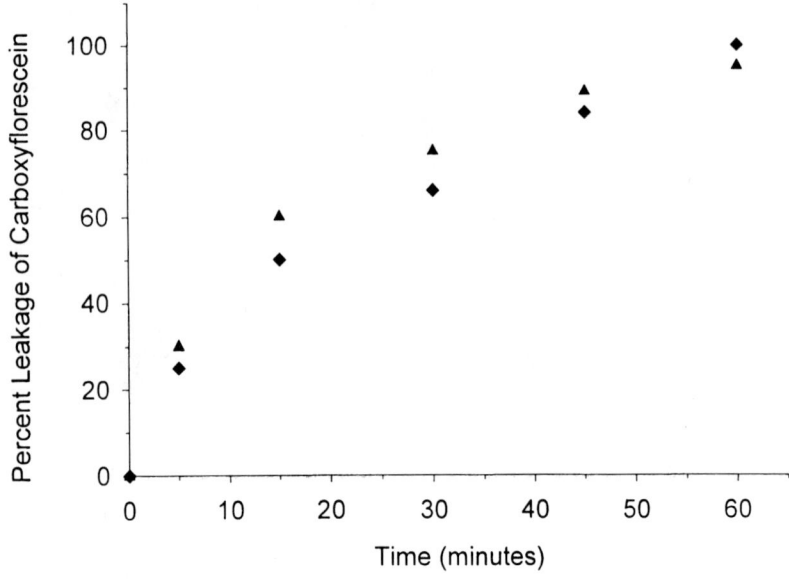

Figure 7. Percent leakage of carboxyfluorescein from DPPE-DVBA vesicles; (▲) non-cross-linked vesicles and (♦) cross-linked vesicles.

with 5 min of UV irradiation. This makes the system suitable for encapsulation of fragile molecules such as enzymes. According to thermotropic results, there is no major change in the membrane properties after cross-linking. TEM images of vesicles before and after cross-linking show no significant change in appearance or size. The cross-linked vesicles can be freeze-dried, redispersed in water and remain smooth, spherical, unilamellar and abundant in large vesicle populations. The structural strength of the cross-linked vesicles has been shown to resist disruption of their three-dimensional structure when exposed to protic and aprotic organic solvents. Non-cross-linked and cross-linked DPPE-DVBA vesicles showed complete release of entrapment of water-soluble carboxyfluorescein in one hour at room temperature. Hydrophobic dye oil red-O was encapsulated in the bilayer of stabilized vesicles after cross-linking. This was supported through lowering of the chain melting transition of the cross-linked vesicles due to an increased disorder of the acyl chains from the visible dye. Filling of cross-linked vesicles with a hydrophobic dye after cross-linking is also a novel application. In addition, the design of similar amphiphiles can be applied as in EGGPE-DVBA in order to facilitate lowering the overall cost in technology development.

References

1. Guo, X.; Szoka Jr., F. C. *Acc. Chem. Res.* **2003**, *36*, 335.
2. Brown, M. D.; Schatzlein, A.; Brownlie, A.; Jack, V.; Wang, W.; Tetley, L.; Gray, A. I.; Uchegbu, I. F. *Bioconjugate Chem.* **2000**, *11*, 880.
3. Nardin, C.; Thoeni, S.; Widmer, J.; Winterhalter, M.; Meier, W. *Chem. Commun.* **2000**, 1433.
4. Graff, A.; Winterhalter, M.; Meier, W. *Langmuir* **2001**, *17*, 919.
5. Walde, P.; Ichikawa, S. *Biomolecular Engineering* **2001**, *18*, 143.
6. Petrikovics, I.; McGuinn, W. D.; Sylvester, D.; Yuzapavik, P.; Jiang, J.; Way, J. L.; Papahadjopoulos, D.; Hong, K.; Yin, R.; Cheng, T.-C.; DeFrank, J. J. *Drug Delivery* **2000**, *7*, 83.
7. Chen, C.-H.; Engel, S. G. *Chem. Phys. Lipids* **1990**, *52*, 179.
8. Juliano, R. L.; Regen, S. L.; Singh, M.; Hsu, M. J.; Singh, A. *Bio/Technol.* **1983**, *1*, 882.
9. Lin, H. Y.; Thomas, J. L. *Langmuir* **2003**, *19*, 1098.
10. Lasic, D. D. *Angew. Chem. Int. Ed. Engl.* **1994**, *33*, 1685.
11. Lasic, D. D.; Needham, D. *Chem. Rev.* **1995**, *95*, 2601.
12. Akimoto, A.; Dorn, K.; Gros, L.; Ringsdorf, H.; Schupp, H. *Angew. Chem. Int. Ed. Engl.* **1981**, *20*, 90.
13. Dorn, K.; Klingbiel, R. T.; Specht, D. P.; Tyminski, P. N.; Ringsdorf, H.; O'Brien, D. F. *J. Am. Chem. Soc.* **1984**, *106*, 1627.
14. Singh, A.; Schnur, J. M. *Phospholipids Handbook*, G. Cevc Ed., Marcel Dekker: New York 1993; pp. 233.
15. Liu, S.; O'Brien, D. F. *Macromolecules* **1999**, *32*, 5519.
16. Lawson, G. E.; Breen, J. J.; Marquez, M.; Singh, A.; Smith, B. D. *Langmuir* **2003**, *19*, 3557.
17. Ravoo, B. J.; Engberts, J. B. F. N. *J. Chem. Soc., Perkin Trans.* **2001**, *2*, 1869.
18. Liu, S.; O'Brien, D. F. *J. Am. Chem. Soc.* **2002**, *124*, 6037.
19. Discher, D. E.; Eisenberg, A. *Science* **2002**, *297*, 967.
20. Discher, B. M.; Bermudez, H.; Hammer, D. A.; Discher, D. E. *J. Phys. Chem. B*, **2002**, *106*, 2848.
21. Katagiri, K.; Hamasaki, R.; Ariga, K.; Kikuchi, J-I. *J. Am. Chem. Soc.* **2002**, *124*, 7892.
22. Rilling, P.; Walter, T.; Pommersheim, R.; Vogt, W. *J. Membr. Sci.* **1997**, *1*, 283.
23. Lawson, G. E.; Lee, Y.; Singh, A. *Langmuir* **2003**, *19*, 6401.
24. Tsukanova, V.; Grainger, D. W.; Salesse, C. *Langmuir* **2002**, *18*, 5539.
25. Mansour, H.; Wang, D-S.; Chen, C-S.; Zografi, G. *Langmuir* **2001**, *17*, 6622.

26. Strzalka, J.; Chen, X.; Moser, C. C.; Dutton, P. L.; Ocko, B. M.; Blasie, J. K. *Langmuir* **2000**, *16*, 10404.
27. Pinazo, A.; Infante, M. R.; Park, S. Y.; Franses, E. I. *Colloid Surface B* **1996**, *8*, 1.
28. Blume, A. *Chem. Phys. Lipids* **1991**, *57*, 253.

Chapter 14

Nanoencapsulation of Organophosphorus Acid Anhydrolase with Mesoporous Materials for Chemical Agent Decontamination in Organic Solvents

Kate K. Ong[1,2], Tu-Chen Cheng[2], Ray Yin[3], Hua Dong[1], Jian-Min Yuan[4], and Yen Wei[1,*]

[1]Department of Chemistry, Drexel University, Philadelphia, PA 19104
[2]Edgewood Chemical and Biological Center, U.S. Army, Aberdeen Proving Ground, MD 21010
[3]ANP Technologies, Inc., 824 Interchange Boulevard, Newark, DE 19711
[4]Department of Physics, Drexel University, Philadelphia, PA 19104

Organophosphorus acid anhydrolase (OPAA) was successfully nanoencapsulated, in-situ, into both silica and organic-modified silica following the low volume shrinkage, non-surfactant templated sol-gel process with D-fructose and poly(ethylene glycol) as the pore-forming agents. By varying the concentration of the template or the concentration of the starting materials, the pore parameters were tuned to have high surface area of 500–800 m^2/g, large pore volume from 0.2–0.8 cm^3/g, and pore diameters ranging from 2–6 nm. As a result, the enzyme remained active in the nanoencapsulated form in both aqueous and mixed aqueous-organic solvents and was reusable by employing a simple regeneration procedure of buffer wash. The immobilization of OPAA in mesoporous materials significantly increased the stability of OPAA against the denaturation in the presence of organic solvents. For a remarkable example, the organically-modified gel sample in 20% acetone significantly retained enzyme activity up to ~90% in comparison with aqueous solution.

Introduction

The development of enzymes operating in extreme conditions has now become the latest scientific quest, especially in the area of industrial biotechnology. One rapidly growing area is bioremediation, or the development of products and techniques for cleaning up pollution caused by agriculture, industry, or urbanization. This particular aspect has become a major concern for the United States military in the safe and effective demilitarization of stockpiled chemical weapons or the decontamination of contaminated areas from chemical or biological agent attacks. In addition, new technologies are needed to rapidly respond to the looming effects from chemical and biological attacks.

In 1980, the U. S. Army initiated a small bioremediation program to identify non-toxic, non-corrosive, environmentally friendly, and effective means for disposal of chemical weapons or neutralizing contaminated areas (*1*). The current decontamination practices were either toxic to both the environment and personnel, or extremely corrosive for usage on sensitive equipment. In either case, large quantities of hazardous waste were generated. Additionally, the weight and volume of the package required large specialized storage facility. These factors combined became a logistical nightmare, especially in non-established facilities, as in the case of deployments. To appease such problems, a group of U.S. Army scientists discovered an enzyme within the *Alteromonas* species that was indigenous to the warm springs of Grantsville, Utah (*2, 3*). This enzyme exhibited good catalytic properties against fluorine containing organophosphorus compounds, including chemical G-type nerve agents. Because of its catalytic properties, this enzyme was classified as organophosphorus acid anhydrolase (OPAA). Several OPAAs from *Alteromonas* strains were screened from which Cheng et al. successfully isolated, cloned, sequenced, and expressed *Alteromonas* sp. strain JD6.5 (*4*). Compared to other organophosphorus hydrolyzing enzymes, this OPAA (*Alteromonas* sp. strain JD6.5) exhibited the best activity against the G-type nerve agents (GD>GF~DFP>GB>GA) with no activity against pesticides and V-agents (*4*). This enzyme was both thermophilic and halophilic–traits important for industrial application, especially for decontamination in harsh environments. Efforts to produce a large quantity of *Alteromonas* sp. strain JD6.5 OPAA, with the goal of developing stabilized enzyme for long-term storage and decontamination, are being investigated (*5, 6*). For example, the enzyme can be packaged as a dry powder form that could be reconstituted with various aqueous solutions for use in existing aqueous-based fire-fighting equipment. These results are promising whereby 60–85% of the enzyme activity retained in the presence of foams, such as fire fighting foams (*6*). In 2004, OPAA was made commercially available and marketed by Genencor International Inc. (Rochester, NY).

Still, scientists are continually conducting research to identify the enzyme's full potential applications. Both Simonian et al. (*7*) and Letant et al. (*8*) demonstrated enzyme immobilization onto silica matrix and porous silicon, respectively, for the development of a nerve agent biosensor. In either case, the enzyme was attached onto a functionalized surface for the detection of fluorophosphate surrogates, DFP and *p*-nitrophenol, commonly-used simulants for G-type nerve agents. Rajan (*9*) explored the feasibility of immobilizing the enzyme onto various fabric materials for the development of an agent protected garment. For in vivo applications, Petrikovics (*10*) focused on the antagonizing organophosphorus intoxication by nanoencapsulating the OPAA enzyme into sterically-stabilized liposomes. Aside from developing a method that retains enzyme activity upon immobilization, another important practical aspect requiring exploration for enzyme-based applications is reusability. The ideal outcome is reduced logistical burden. The nanoencapsulated enzyme would provide less material with minimal storage and operational cost as compared to free enzyme.

Enzyme encapsulation into mesoporous molecular sieves shows great promise. Wei's laboratory has developed a novel non-surfactant templated approach (*11–16*) that is both biologically and environmentally friendly as well as easy to produce. The materials are based on common silicates or other metal oxides. The added bonus of using a solid matrix is that the material as biocatalyst can be reused. With this, successful encapsulation of enzymes (*13–16*) such as horseradish peroxidase, alkaline phosphatase, and glucose oxidase has been demonstrated. In all cases, the enzyme remained active after encapsulation. It is believed that the silicate material forms a protective cage that physically prevents the enzyme from completely unfolding its 3-dimensional structure, an, yet, the cage allows space for the conformational reactivity, thereby retaining enzyme activity. Depending on the sample to be nanoencapsulated, the pore (cage) size and chemistry contained within mesoporous material can be tailored. No enzyme modification is required. This is extremely important, especially when using rare enzymes. In this study, we will attempt to encapsulate OPAA into mesoporous type material following the non-surfactant templated approach. These materials will be tested for enzyme activity under aqueous buffer conditions as well as under mixed organic/aqueous media as found in contaminants commonly found in field environment. To date, there has been no systematic research on nanoencapsulated enzyme materials prepared by the non-surfactant templated approach in solutions containing organic solvents. In addition, the reusability aspect of the prepared materials will be investigated.

Materials and Experimental Procedures

Materials

Precursors, TMOS (tetramethyl orthosilicate) and MTMS (methyltrimethoxysilane), were purchased from Aldrich Chemical Company (Milwaukee, WI). Diisopropyl fluorophosphates (DFP), 1,3-bis(trishydroxymethylmethylamino) propane (BTP), D-fructose, polyethylene glycol (PEG, MW 400), hydrochloric acid (HCl), dimethyl formamide (DMF), acetone, methanol, and hexane were purchased from Sigma Chemical Company (St. Louis, MO). Sodium fluoride was purchased from Fisher Scientific (Fair Lawn, NJ). Organophosphate acid anhydrolase (OPAA, EC 3.1.8.2) was purified and provided by Dr. Tu-chen Cheng (U.S. Army). All chemicals and reagents were used as received.

Mesoporous Material Preparation

The preparation of OPAA entrapped mesoporous silica materials followed procedures as previously described (*11–17*). In general, a low volume shrinkage sol-gel procedure was developed by our laboratory to immobilize sensitive biomolecules into the non-surfactant templated mesoporous materials. Both fructose and PEG were used as templates whereby both TMOS and mixtures of TMOS/MTMS (at molar ratios of 3:1, 1:3, and 1:1) were used as the precursors for the pure silica and organically-modified silica materials. The precursors were hydrolyzed at room temperature by mixing in a reaction vessel the precursor(s), distilled water, and HCl at room temperature. Because this hydrolytic reaction was exothermic, the temperature of the solution elevated. The acid-hydrolyzed precursors were degassed under vacuum to remove the presence of by-products, i.e., water and methanol, which were collected into a cold-trap. Degassing continued until the formation of bubbles in the solution slowed until it required several minutes for a single bubble to form. This degassing step was important to eliminate the denaturing effect of methanol on the enzyme. After the vacuum evacuation, the system became very viscous. The pH of the mixture was adjusted to 7.3 with 50 mM bis-tris propane buffer prior to OPAA addition. The aqueous template solution followed by 0.5 mL OPAA enzyme solution (0.30 mg in 50 mM bis-tris propane buffer pH 7.3) was added to complete the sol mixture and the system was allowed to gel at room temperature. The overall OPAA load per sample was 0.30 mg. The reaction vessel was covered with Paraffin film and pierced with a syringe needle to allow the condensation and evaporation of the solvent and any by-products (i.e., water and methanol) to continue overnight. A vacuum oven at room temperature was used to further evacuate and dry the gel until a constant weight was reached, whereby less than 5% change in weight occurred in a 24 h period. The final products were transparent, millimeter-sized, biogel disks, which were crushed with a mortar and pestle into a fine powder and stored in a $-15\,°C$ freezer. Table I summarizes the preparation and pore parameters of various OPAA entrapped samples.

OPAA Activity Determination

Activity measurements were performed following procedures similar to those elsewhere in the literature (5). In general, the amount of entrapped-enzyme polymers required for a 10 μg OPAA preparation was calculated based on the total mass of the polymer and the enzyme loading. The nanoencapsulated-enzyme polymers, as summarized in Table I, were weighed to contain approximately 10 μg OPAA, and transferred into a 15-mL centrifuge tube. For example, a 10 μg OPAA sample for the TMOS-F50-OPAA preparation, weighing 1.3967 g with an enzyme loading of 0.3 mg required 94.14 mg of the TMOS-F50-OPAA preparation. Prior to analysis, the template for the nanoencapsulated-enzyme polymers was extracted by adding 15 mL 10 mM bis-tris propane buffer, pH 7.2, to each centrifuge tube sample. After rocking for 1 h in a refrigerator at 6 °C, the samples were centrifuged at 3200 rpm for 10 min. Supernatant was decanted and retained for analysis of enzyme leakage. This extraction process was repeated two more times. The extracted sample was then stored in its centrifuge tube at 6 °C until use.

For all activity analyses, the measurement proceeded in a thermally-controlled, water-jacketed, reaction vessel. Note, all solutions including the nanoencapsulated enzyme were equilibrated to 25.0 °C prior to analysis. 2.5 mL of 50 mM bis-tris propane containing 1 mM $MnCl_2$, pH 8.5, was pipetted into the centrifuge tube containing the extracted enzyme sample. The sample was mixed and transferred into the reaction vessel, equipped with a magnetic stirrer and a fluoride sensitive electrode. A final concentration of 0.3 mM DFP was then added to initiate the enzyme-substrate reaction. One unit of enzymatic activity was based on the release of 1 micromole of fluoride per minute at 25 °C. To determine the sample volume for the activity determination of the solid nanomaterials, a density function was incorporated. Therefore, samples containing OPAA and pore-former utilized 1500 g/mL, while the control samples and the wet gel samples used 2300 g/mL. As a result, the sample volume for the activity determination was simply equated by dividing the measured sample weight by the density. Specific activity was expressed as units per mg of enzyme, after correction from spontaneous hydrolysis in the absence of OPAA under otherwise identical conditions using the following equation (2–4):

Table I. Properties of OPAA containing Mesoporous Silica Materials Synthesized from the Low Volume Shrinkage Process Approach

Matrix	Precursor (molar ratio)	Template	Weight for 10 mg OPAA[a] (mg)	Surface Area[b] (m^2/g)	Pore Diameter[c] (Å)	Pore Volume[d] (cm^3/g)
TMOS-F50-OPAA	TMOS	50% Fructose	94.14	717	40	0.73
TMOS-PEG-OPAA	TMOS	50% PEG	91.73	723	32	0.71
MTMS-PEG-OPAA	TMOS/MTMS (1:1)	50% PEG	86.67	744	<15	0.61
3MTMS-F50-OPAA	TMOS/MTMS (1:3)	50% Fructose	86.77	776	33	0.673
3MTMS-wet gel	TMOS/MTMS (1:3)	50% PEG	123.78	---	>50	---

[a] Calculated based on the prepared weight of the material and the enzyme loading (0.3 mg)
[b] Determined by the BET method (17)
[c] Determined by the t-plot method (17)
[d] Determined at $P/P_o=1$ (17)
[e] free, native OPAA = 1202 U/mg
--- indicate data unavailable

$$R = C_\Delta - C_{sub} \left[\frac{E_1}{V_1 xT} \right]$$

whereby R = micromoles/min/mL; C_Δ = C_{6min} - C_{2min} in μM measured from ISE meter; C_{sub} = DFP spontaneous rate; E_1 = 1/E; E = sample volume in mL; V_1 = 1000/V; V = total assay volume in mL = 2.5; and T = time of assay in min = 4.

After each activity measurement, the nanoencapsulated-enzyme samples were transferred with a pipet from the reaction vessel back to a centrifuge tube for reusability analyses. The samples were centrifuge washed, as described earlier, for three times, in 10 mM bis-tris propane, pH 7.2, and the supernatant was decanted and retained for analysis of enzyme leakage. The final washed sample was then stored, decanted dry, in its tube at 6 °C until next trial.

Results and Discussion

Alteromonas sp. JD6.5 OPAA has been shown (4–6) to be extremely important for its ability to detoxify chemical nerve agents. For this reason, scientists are feverishly searching for a means to incorporate this unique enzyme into a platform for either the detection or the detoxification of the toxic organophosphorus compounds. The nanoencapsulation, or entrapment, of the enzyme to reduce logistical burden is a strong driving force.

Mesoporous materials are known for their large pore volume and surface area, as well as controlled pore size. Wei's laboratory has been perfecting the in-situ nanoencapsulation of enzymes into mesoporous materials via the non-surfactant pathway. For horseradish peroxidase (HRP), Dong et al. (13–17) found that the activity of HRP following the non-surfactant pathway was two orders of magnitude higher than those measured by conventional sol-gel immobilization. They also report that the HRP-containing materials exhibit excellent thermal stability and minimal enzyme leakage. Despite this inherent success, the procedure still requires individualized tailoring because each enzyme is different both physically and chemically (such as hydrophobicity, size, folding parameters). For example, lipase (17), though without optimization, has resulted in poor enzyme activity when nanoencapsulated using the same process as used for HRP nanoencapsulation. Likewise, entrapment of the well-characterized organophosphorus hydrolase (OPH, another organophosphorus detoxifying enzyme) has been studied whereby Lei et al. (18) electrostatically immobilized OPH onto the functionalized surface wall of pre-synthesized mesoporous material. Unlike Lei et al.'s method, Wei's method is much simpler to perform whereby the enzyme and material are formed in one-step, without the need for post-functionalization. Additionally, the enzyme is encapsulated and protected within the mesoporous cage, rather than being fully exposed to the potentially harsh environment; this results in its enhanced stability.

Following Wei's approach, the traditional, non-surfactant templated process was modified to include a low volume shrinkage process, which involved degassing and buffering of the hydrolyzed sol prior to addition of OPAA since the traditional method yielded samples that exhibited very low activity (data not shown). As the sample dried from the time it was prepared, the activity decreased (Figure 1). Each component of the sample preparation process was analyzed, and it was found that the presence of methanol, a by-product of the sol hydrolysis of the silicate materials, affected OPAA activity (data not shown) by possibly denaturing the enzyme. Hence, the procedure was modified to minimize this denaturing effect and be more biocompatible.

Utilizing the low volume shrinkage process, a total of five samples of nanoencapsulated OPAA were prepared along with control samples. The control samples either lacked the pore-forming agent (template) or the enzyme. Note that data for the control samples are not presented. Table I summarized the physical pore parameters (pore size, surface area, and pore volume) and

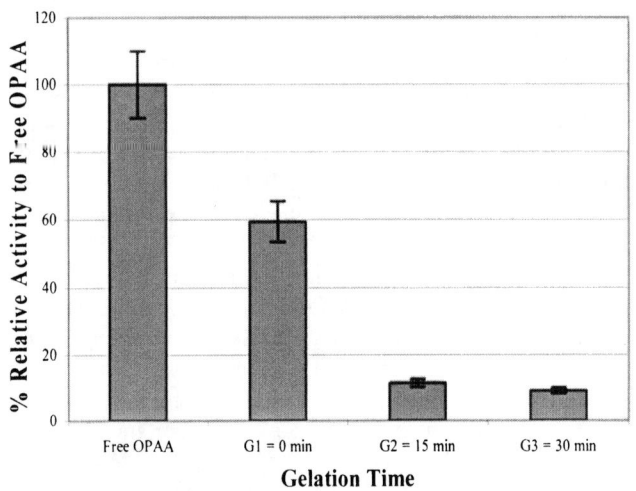

Figure 1. Percent relative enzymatic activity of the nanoencapsulated preparation gelation time to free OPAA in 50 mM bis-tris propane, pH 8. The nanoencapsulated preparation contained TMOS + MTMS + OPAA. The activity was measured at initial mixture (G1 = 0 min), 15 min later (G2 = 15 min), and 30 min later (G3 = 30 min). Standard condition was 50 mM bis-tris propane, pH 8. Free enzyme in standard condition is 1013 U/mg.

conditions for each of the nanoencapsulated OPAA samples as prepared using either D-fructose or PEG as the pore forming agent or template. These physical parameters were previously analyzed (17) following the Brunauer-Emmett-Teller (BET), nitrogen adsorption-desorption isotherms, and the Barrett-Joyner-Halenda models. Overall, three samples (pure silica matrices: TMOS-F50-OPAA and TMOS-PEG-OPAA and the organic modified matrix: 3MTMS-F50-OPAA) exhibited mesoporosity, i.e., narrowly distributed pore size (3–4 nm), large surface area (~700 m^2/g), and large pore volume (0.6–0.7 cm^3/g). Though the additional methyl group in the MTMS-PEG-OPAA sample increased the hydrophobicity, it also decreased the pore volume from the pure silica TMOS samples, making a predominately microporous sample. Likewise, the 3MTMS-F50-OPAA was prepared with 3 molar ratio of methyl group, making this sample more hydrophobic, but with better pore volume. However, this sample appeared opaque. Dong (17) postulated a phase separation phenomena whereby the increase in methyl concentration caused the fructose to precipitate from the hydrolyzed sol. This effect is minimized using PEG to accelerate the gelation and enable faster immobilization of enzyme, especially in the organic modified samples. In comparison, the 3MTMS-wet gel is an incomplete dried sample of 3MTMS-F50-OPAA and was predominantly macroporous. For practical applications, completely dried samples are preferred for they offer better stability (a rigid mesoporous cage) and reusability, which are difficult to control in the partially dried (wet) gel.

Aqueous Solution

The activities of each nanoencapsulated OPAA and its reusability results in 100% aqueous buffer conditions are summarized in Table II. The 100% aqueous condition or the standard buffer condition was 50 mM bis-tris propane, pH 8. No significant difference was noted in the TMOS sample between use of PEG or D-fructose as the template. When methyl group was incorporated into the silica matrix as in MTMS-PEG-OPAA, a slightly higher activity (<1%) was measured compared to the pure silica matrix. Also, as the molar concentration of methyl increased, the enzyme activity decreased significantly (5 times), as seen in the 3MTMS-F50-OPAA sample. Two possible reasons were suggested. One, as methyl concentration increased, the gelation time increased, thereby prolonging the mentioned "methanol effect" causing the enzyme to denature. Two, fructose precipitates from the hydrolyzed sol at high concentration of organosilicate, resulting in an inhomogeneous opaque sample. Quite the opposite was observed for the partially hydrolyzed 3MTMS-wet gel sample. A very high activity was measured with this wet biogel, nearly 50% comparable to free OPAA. Because this sample was more of a liquid wet gel rather than a hard, dry sample, perhaps it provided a more OPAA-friendly environment or

Table II. Summary of OPAA Activity from the Reusability Test Data for each Nanoencapsulated Sample Matrix in Aqueous Solution

Reuse Trial	Sample Matrix				
	TMOS-F50-OPAA	TMOS-PEG-OPAA	MTMS-PEG-OPAA	3MTMS-F50-OPAA	3MTMS-wet gel
Initial	294.4 ± 29.9	285 ± 84.2	368.5 ± 80.2	69.9 ± 30.5	605.4 ± 85.3
1	547.5 ± 65.7	483.9 ± 140.3	487.9 ± 102.4	5.6 ± 2.2	365.5 ± 51.2
2	584.6 ± 70.1	348.7 ± 97.6	332.2 ± 69.7	5.9 ± 2.4	290.1 ± 40.6
3	787.8 ± 94.5	512.8 ± 143.5	287.7 ± 60.4	0	143.0 ± 20.1
4	549.5 ± 60.4	353.2 ± 98.9	247.5 ± 51.9	0	59.0 ± 8.3
5	452.3 ± 54.3	228.1 ± 63.8	375.9 ± 78.9	0	0
6	533.6 ± 53.3	329.4 ± 92.2	390.5 ± 82.0	0	0
7	388.7 ± 46.6	313.9 ± 87.8	193.2 ± 40.5	0	0
8	254.9 ± 30.6	240.1 ± 48.0	133.3 ± 27.9	0	0

NOTE: Activities are in units per milligram (U/mg)

easier access to enzyme by substrate. Unlike the 3MTMS-F50-OPAA sample, this wet gel appeared clear, with no signs of fructose precipitation.

To demonstrate that no special equipment and chemicals were required, a very simple regeneration wash procedure was used for the reusability study (Table II). After each activity measurement, these samples were simply washed with a 10 mM bis-tris propane buffer at pH 7 and stored, decanted, until the next measurements. Of all the matrices, 3MTMS-F50-OPAA performed the worst. Even though this sample is a phase-separated material, it could be reused three times before total activity was lost. Perhaps the enzyme-active site was not completely blocked by the precipitated fructose, thereby allowing for the enzyme-substrate interaction. Still, this result is much better than using free, untrapped enzyme, where there is no reusability. Compared to the dry matrices, the wet 3MTMS-wet gel could only be reused about four times before 90% of its activity was lost. This was not surprising considering its macroporous, fluid-like nature. Likewise, reusability studies with TMOS and MTMS samples (using PEG as the pore-former) were randomly tested over a 75-day period and could be reused at a minimum 8 times without significant (<50%) loss in activity. Such observations were also noted for the TMOS-F50 sample with fructose as the pore-forming agent. Though not fully investigated, this 75 day time period indicates that long-term storage of the "dry" nanoencapsulated enzyme is possible with very minimal buffer content. The relative activity in both the TMOS and MTMS samples did appear to fluctuate. In some cases, the activity increased after initial use. Possible reasons for fluctuation include (1) loss of sample during wash, (2) OPAA availability to substrate, or (3) fluoride probe performance. Also, high pH has a deleterious effect on the silica structure, perhaps causing the nanoencapsulated OPAA to be exposed and possibly denatured. Such pH phenomenon has been known (*19, 20*) for silica-based supports.

Mixed Aqueous-Organic Solution

To investigate the effects the nanoencapsulated OPAA samples had in mixed aqueous-organic solvent media, the following solvents, arranged in order of increasing polarity (hexane, acetone, methanol and dimethylformamide), were tested at 20% (v/v) in 50 mM bis-tris propane, pH 8. These solvents were chosen for various reasons. (1) They are commercially available and commonly used, (2) they represent a range of solvent properties, and (3) they are components of contaminants commonly found in the field environment. For comparison, the relative activities of free, non-nanoencapsulated OPAA against the various organic solvents are diagrammed in Figure 2 and summarized in Table III. Because of high spontaneous hydrolysis rate of the substrate (DFP) in the organic solvents, the maximum solvent concentration that could be tested

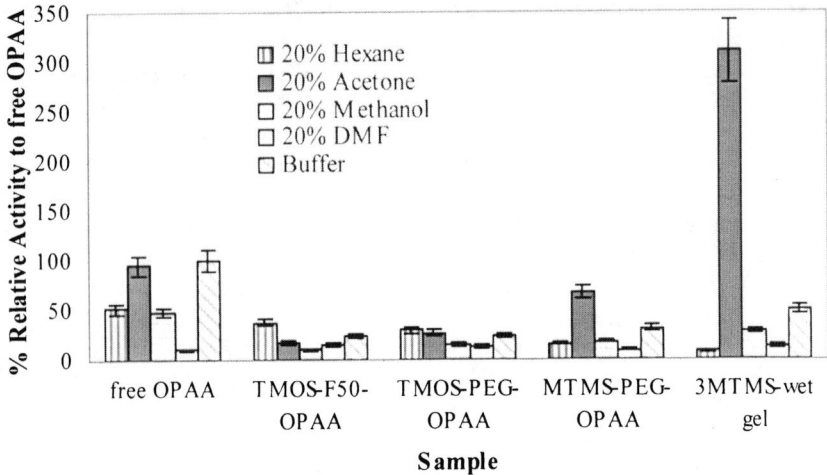

Figure 2. Comparison of the nanoencapsulated OPAA activity in different solvents relative to free OPAA in 50 mM bis-tris propane, pH 8. The nanoencapsulated OPAA samples were TMOS-F50-OPAA, TMOS-PEG-OPAA, MTMS-PEG-OPAA, 3MTMS-F50-OPAA, 3MTMS-wet gel, 3MTMS-OPAA, and MTMS-OPAA. The solvents studied were 20% hexane (vertical bars), 20% acetone (shaded bars), 20% methanol (horizontal bars), 20% dimethylformamide (unshaded bars), and control buffer, 50 mM bis-tris propane (diagonal bars). The activity of free OPAA in 50 mM bis-tris propane, pH 8 was 1202 U/mg.

with the fluoride ion-selective electrode was 20%. At higher organic solvent concentrations, the fluoride probe performance was greatly affected.

Exciting results are observed in Figure 2 or Table III. First and foremost, excellent enzyme activity is observed in the various mixed aqueous-organic solvents, especially in the nanoencapsulated OPAA samples. This could be attributed to the fact that OPAA molecules are nanoencapsulated within nanoscale-confined space of the mesoporous matrix. Because of the space confinement, the protein unfolding (i.e., denaturing), as traditionally caused by organic solvents, is significantly restricted; thereby, significant retention of enzyme activity is achieved (17). This denaturing effect of free OPAA was witnessed by the loss in relative activity for the methanol (~50% loss), hexane (~50% loss), and dimethyl formamide (88% loss) tested samples. Secondly, Figure 2 appears skewed because the OPAA nanoencapsulated 3MTMS-wet gel sample in 20% acetone demonstrates extremely high activity, with a 210%

Table III. Nanoencapsulated OPAA Activity in Various Solvents

Matrix	Solvent				
	100% Aqueous	20% Hexane	20% DMF	20% Acetone	20% Methanol
free OPAA	1202.08 ± 177.34	619.01 ± 51.93	124.48 ± 13.69	1109.20 ± 115.64	579.33 ± 63.72
TMOS-F50-OPAA	294.25 ± 29.93	455.30 ± 40.98	183.78 ± 20.22	204.98 ± 22.55	127.05 ± 13.98
TMOS-PEG-OPAA	285.02 ± 84.24	358.19 ± 39.40	158.56 ± 17.44	321.09 ± 35.32	179.80 ± 19.78
MTMS-PEG-OPAA	368.55 ± 80.23	192.37 ± 21.16	113.21 ± 12.45	808.16 ± 88.90	217.82 ± 23.96
3MTMS-wet gel	605.39 ± 85.28	96.78 ± 10.65	159.43 ± 18.33	3726.97 ± 409.97	341.24 ± 37.53
3MTMS-F50-OPAA	69.82 ± 30.48	47.49 ± 6.22	11.78 ± 2.33	114.35 ± 12.58	0.00

NOTE: Activities are in Units per milligram (U/mg)

enhancement when compared to free OPAA in standard buffer condition. As suggested by Baek et al. (*21*), the acetone enhanced enzyme activity by creating a hydrophobic environment in the active site. This, combined with the hydrophobic and fluidic nature of the 3MTMS-hydrogel sample, creates an environment amenable for the hydrophobic OPAA. Though not as dramatic, this acetone enhancement effect was also observed in the TMOS-PEG-OPAA, 3MTMS-F50-OPAA, and MTMS-PEG-OPAA sample. This effect is interesting because acetone is commonly used for purification of protein samples by lowering the protein's solubility, thereby causing the proteins to precipitate. Either the acetone further purified the OPAA sample from any remaining cellular impurities that may have affected the enzyme-substrate interaction, or perhaps, the acetone along with the 3MTMS-wet gel matrix appears to have an enhancing effect on the enzyme unlike any other sample. Third, good enzyme activity was measured for the nanoencapsulated OPAA in 20% hexane, compared to denaturing effect of the sample to the free enzyme. This is especially noted in the TMOS-F50-OPAA and the TMOS-PEG-OPAA samples, whereby the hexane sample exceeded the activity measured in all solvents. Fourth, virtually no enzyme activity was detected in the control samples prepared without the pore-forming templates. This result demonstrated the need for pore-forming templates to enable enzyme-substrate accessibility.

The protective advantage the matrix has on the enzyme is particularly evidenced in the dimethyl formamide (DMF) samples, as shown in Figure 3. As mentioned above (Figure 2), the free OPAA exhibited the lowest activity in the presence of DMF (88% loss) relative to free OPAA, also in 20% DMF. Once the enzyme was nanoencapsulated, enzyme activity increased by about 40% compared to that of free OPAA, especially in both TMOS and 3MTMS-wet gel samples. Unfortunately, the enzymatic activity in the 3MTMS-F50-OPAA is still very low. Perhaps the polar nature of the DMF solvent which contained the substrate limited its ability to interact with the hydrophobic OPAA nanoencapsulated within the hydrophobic 3MTMS-F50-OPAA.

Earlier, it was determined that methanol had a deleterious effect on the free enzyme; hence, the low shrinkage approach was employed. Figure 4 shows a plot of the enzyme activity of the nanoencapsulated OPAA relative to free OPAA in 20% methanol solution. Though the overall activity of the nanoencapsulated samples decreased by about 70% from its free format, enzyme activity was measured in all matrices, except for 3MTMS-F50-OPAA, which exhibited no detectable activity. Perhaps, the additional methyl groups were too hydrophobic, thereby hindering the enzymatic activity. Also, the microporous nature of this sample can limit the enzyme-substrate interaction. In the case of 3MTMS-wet gel, this wet gel is physically more fluidic and more amenable toward maintaining enzymatic activity, as well as being less restrictive for the enzyme-substrate interaction.

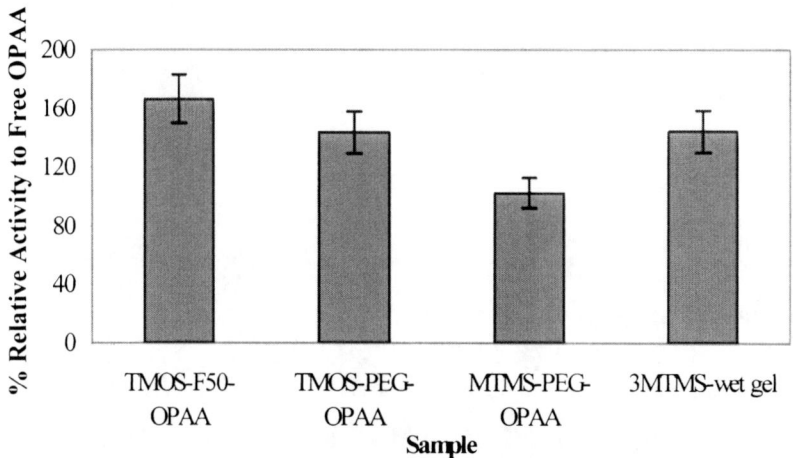

Figure 3. Percent relative activity of various entrapped OPAA samples against free OPAA in 20% dimethyl formamide. The matrices were TMOS-F50-OPAA, TMOS-PEG-OPAA, MTMS-PEG-OPAA, and 3MTMS-wet gel. Free OPAA activity in 20% dimethyl formamide was 124 U/mg.

Reusability

Aside from determining whether the nanoencapsulated enzyme retains activity in the presence of organic solvent, another important feature for nanoencapsulation is its potential for reuse. Table IV summarizes the activity measurements of the reused nanoencapsulated OPAA samples in the various mixed aqueous-organic solvent solutions (methanol, acetone, dimethyl formamide, and hexane) at 20% (v/v) in 50 mM bis-tris propane, pH 8. From these data, it became apparent 3MTMS-hydrogel is not suitable for reusability, regardless of solvent, with about 80% of its activity lost after initial use. Though, this sample provides a suitable environment for the enzyme, it lacks the physical rigidity that the other matrices possess necessary for protecting the enzyme. In comparison, the pure silica matrices (TMOS-F50-OPAA and TMOS-PEG-OPAA) and the organically modified matrix (MTMS-PEG-OPAA) demonstrated reusability at least two times before complete loss of activity.

Some of the activity loss may be reflected by the loss of sample during the wash and transfer process. For the TMOS samples, the enzyme activity actually increased from the first reuse in the tested solvents, except for DMF. The increase for the TMOS samples prepared in D-fructose was twice the initial reading, while it was slightly less for the TMOS-PEG samples. As mentioned earlier, the polar nature of DMF probably limited the interaction between the

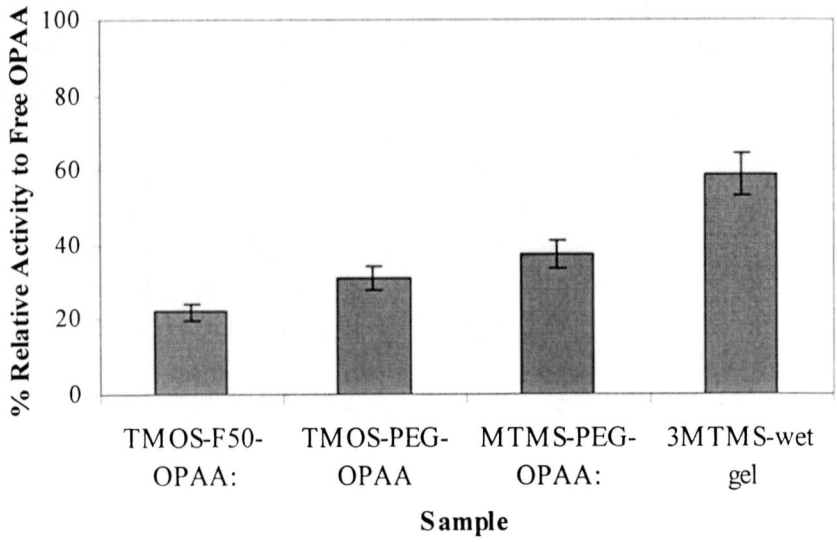

Figure 4. Percent relative activity of various nanoencapsulated OPAA samples against free OPAA in 20% methanol. The matrices were TMOS-F50-OPAA, TMOS-PEG-OPAA, MTMS-PEG-OPAA, and 3MTMS-wet gel. Free OPAA activity in 20% methanol was 579 U/mg.

hydrophobic enzyme and the substrate. In comparison, the enzyme activity from the methylated (MTMS or 3MTMS) hybrid materials generally decreased with each reuse. Several reasons may account for these fluctuations in activity after initial use. Because template extraction was performed in buffer, some of the template might not be removed. The presence of organic solvent might help clean the mesopores by removing the template and allowing easier enzyme-substrate accessibility. The mixed organic solvent buffer may also react with the physical structure of the mesoporous material, thereby exposing the enzyme. It has been reported that silica-based supports are not stable in alkaline medium or by prolonged use in neutral pH, aqueous solutions, thereby causing the immobilized dopant to leak from the support (*19, 20*). Because the OPAA activity was measured at pH 8.5, exposure at this pH or remnants of test solvent (which contains the substrate) left in the sample caused the matrix to slowly change over time. However, no activity was detected by the fluoride ion-selective electrode in the supernatant from the centrifuge washes.

For all materials, minimal (<1%) to no enzyme activity was measured by the fluoride ion selective electrode in the control samples (materials without OPAA and materials without pore-forming agent, i.e., D-fructose or PEG) as well as in the supernatants from the wash steps (data not shown). The result for

Table IV. Reusability Activity Summary of the Free and Entrapped OPAA in Various Aqueous-Organic Solvent

Solvent (20% v/v)	Trial	Free OPAA	TMOS-F50-OPAA	TMOS-PEG-OPAA	MTMS-PEG-OPAA	3MTMS-wet gel	3MTMS-F50-OPAA
Hexane	initial	619.01 ± 51.93	455.30 ± 40.98	358.20 ± 39.40	192.37 ± 21.16	96.78 ± 10.65	47.49 ± 6.22
	reuse 1		821.41 ± 98.56	555.74 ± 66.69	111.81 ± 12.30	61.59 ± 7.39	17.93 ± 2.15
Methanol	initial	579.33 ± 63.72	127.05 ± 13.98	179.80 ± 19.78	217.82 ± 23.96	341.24 ± 18.33	0.00
	reuse 1		255.69 ± 30.68	216.79 ± 26.01	196.31 ± 21.59	37.33 ± 4.48	0.00
	reuse 2		141.66 ± 16.99	222.41 ± 26.69	281.72 ± 33.81	0.00	0.00
	reuse 3		230.54 ± 27.67	0.00	0.00	0.00	0.00
DMF	initial	124.48 ± 13.69	183.78 ± 20.22	158.56 ± 17.44	113.21 ± 12.45	159.43 ± 18.33	11.78 ± 2.33
	reuse 1		61.12 ± 7.33	82.83 ± 9.94	70.65 ± 8.50	45.73 ± 5.49	0.00
Acetone	initial	1109.20 ± 115.64	204.98 ± 22.55	321.09 ± 35.32	808.16 ± 88.90	3726.97 ± 409.97	114.35 ± 12.58
	reuse 1		407.04 ± 48.93	403.65 ± 48.44	478.47 ± 57.45	114.35 ± 13.72	1884.70 ± 226.16
	reuse 2		609.11 ± 73.09	486.22 ± 58.35	148.79 ± 17.85	42.44 ± 0.05	0.00

NOTE: Activities are in units per milligram (U/mg).

materials without OPAA (i.e., TMOS, 3-MTMS or MTMS) demonstrated that the material itself did not contribute to the overall calculation of enzyme activity. Likewise, the low enzymatic activity in the materials in the absence of the, D-Fructose template, i.e., TMOS-OPAA, 3-MTMS-OPAA, and MTMS-OPAA, demonstrated the importance of the mesoporosity for the enzyme-substrate accessibility. Without the template or pore-forming agent, the material is essentially microporous (15 Å). In other words, the OPAA enzyme was trapped inside the silica matrix with very small channels to outside environment and the diffusion of substrate to the enzyme was limited. Finally, the test of the supernatant from the water extraction or wash step indicated that enzyme leaking is not measurable and insignificant.

Conclusions

Organophosphorus acid anhydrolase (OPAA) has received considerable attention for its potential application in enzymatic decontamination of G-type chemical warfare agents. To improve its use for special and practical applications, mesoporous materials containing OPAA were successfully prepared and analyzed for enzymatic activity and reusability. The nanoencapsulated OPAA samples were prepared following the established non-surfactant templated pathway. Though successful for the entrapment of enzymes such as horseradish peroxidase, the sample preparation procedure was further modified by including an evacuation and buffering step to minimize the denaturing effects of methanol (a by-product of the sol-gel reactions) on the OPAA enzyme. In essence, this newly developed procedure, coined low volume shrinkage approach, was relatively simple and inexpensive to perform in one-pot, with standard laboratory equipment. By varying the concentration of the pore-forming agent or the concentration of the starting materials, the pore parameters were overall mesoporous, exhibiting high surface area of 500–800 m^2/g, large pore volume from 0.2–0.8 cm^3/g, and designable pore diameters ranging from 2–6 nm. Two classes of mesoporous materials were prepared: pure silica using tetramethyl orthosilicate and organically modified adding methyl trimethoxysilicate.

Good OPAA activity was measured with the prepared materials in 100% aqueous buffer solution. Because OPAA is hydrophobic in nature, organic modification of the silicate sample (3MTMS or MTMS) resulted in a slightly higher activity than the pure silicate sample. However, as the molar concentration of the methyl group increased, precipitation of the template, D-fructose, resulted in a relatively poorly active sample when dried. Remarkably good enzymatic activity was found with the partially hydrolyzed wet biogel, which did not show signs of template precipitation. With a simple regenerating wash step, the entrapped OPAA overall demonstrated reusability up to eight times.

Exciting results were obtained with the prepared materials in mixed aqueous-organic solvent solution. Never before had mesoporous materials prepared by the non-surfactant templated pathway been tested in an organic environment. Significant loss in free enzyme activity was measured in 20% (v/v) dimethyl formamide. Once entrapped, enzyme activity increased and was reusable, thereby demonstrating the protective nature that the mesoporous cage has on the stability of the enzyme. In the 20% (v/v) acetone analyses, the activity of the entrapped enzyme was significantly enhanced in all matrices. The greatest enhancement was measured in the wet biogel sample at 210%. Perhaps the acetone removed possible impurities that may affect the enzyme-substrate interaction or the acetone created a hydrophobic environment that was amenable to the hydrophobic OPAA. With the exception of the precipitated 3MTMS-F50-OPAA sample in 20% (v/v) methanol and in 20% (v/v) dimethyl formamide, all samples exhibited enzymatic activity and reusability with little to no enzyme leakage. Finally, little to no activity was measured in the control samples prepared without the pore-forming agents (PEG or D-fructose) or without the enzyme.

Though the goal of nanoencapsulating OPAA into a stable, active, and reusable material was accomplished, more research is still needed. For example, an alternative, non-destructive, analytical method for determining OPAA activity would be helpful to reduce assay time and not be limited by concentration effects. Additionally, a range of solvents would test the limits and stability of the nanoencapsulated OPAA. Other operating parameters, such as pH and temperature, also merit further study.

Acknowledgments

This work was supported in part by the U. S. Army Research Office (ARO), the National Institute of Health (Grant No. DE09848), and the Nanotechnology Institute of Southern Pennsylvania. Sincere gratitude was owed to Dr. Zhengfei Sun for his assistance with the CD analysis of the OPAA samples.

References

1. DeFrank, J.; Cheng, T.; Harvey, S; Rastogi, V. *Proc. of the 2002 Joint Service Sci. Conf. on Chem. & Biol. Def. Research*, 19-21 November 2002, AD-M001 523.
2. DeFrank, J.; Beaudry, W.; Cheng, T.; Harvey, S.; Stroup, A.; Szafraniec, L. *Chem. Biol. Interact.* **1993**, *87*, 141.
3. DeFrank, J.; Cheng, T. *J. Bacteriol.* **1991**, *173*, 1938.
4. Cheng, T.; Harvey, S.; Chen, G. *Appl. Environ. Microbiol.* **1996**, *62*, 1636.

5. Cheng, T.; Rastogi, V.; DeFrank, J.; Sawiris, G. *Enzyme Eng. XIV Annals of the NY Acad. Sci.* **1998**, *864*, 253.
6. Cheng, T.; Calomiris, J. *Proc. of the 1994 ERDEC Sci. Conf. on Chem. Def. Research*, 15–18 November 1994, AD-A313 080.
7. Simonian, A.; Grimsley, J.; Flounders, A.; Schoeniger, J.; Cheng, T.; DeFrank, J.; Wild, J. *Anal. Chim. Acta.* **2001**, *442*, 15.
8. Letant, S. E.; Kane, S. R.; Hart, B. R.; Hadi, M. Z.; Cheng, T. C.; Rastogi, V. K.; Reynolds, J. G. *Chem. Commun.* **2005**, *7*, 851.
9. Rajan, K. *Def. Tech. Info. Ctr.* **1991**, AD-A231-056.
10. Petrikovics, I; Cheng, T.; Papahadjopoulos, D.; Hong, K.; Yin, R.; DeFrank, J.; Jaing, J.; Song, Z.; McGuinn, W.; Sylvester, D.; Pei, L.; Madec, J.; Tamulinas, C.; Jaszberenyi, J.; Barcza, T.; Way, J. *Toxicol. Sci.* **2000**, *57*, 16.
11. Wei, Y.; Oiu, K-Y. Series on Chemical Engineering, 2004, *4* (Nanoporous Materials), 873.
12. Wei, Y.; Jin, D.; Ding, T.; Shih, W.; Liu, X.; Cheng, S.; Fu, Q. *Adv. Mater.* **1998**, *10*, 313.
13. Wei, Y.; Dong, H.; Xu, J.; Feng, Q. *ChemPhysChem*, **2002**, *3*, 802.
14. Wei, Y.; Xu, J.; Dong, H.; Feng, Q. U.S. Patent No. 6,989,254, **2006**.
15. Wei, Y.; Xu, J.; Feng, Q.; Dong, H.; Lin, M. *Mater. Letters* **2000**, *44*, 6.
16. Wei, Y.; Xu, J.; Feng, Q.; Lin, M.; Dong, H.; Zhang, W.; Wang, C. *J. Nanosci. Nanotechnol.* **2001**, *1*, 83.
17. Dong, H. PhD dissertation, Drexel University, Philadelphia, PA, 2002.
18. Lei, C.; Shin, Y.; Liu, J.; Ackerman, E. *J. Amer. Chem. Soc.* **2002**, *124* 11242.
19. Bhatia, R.; Brinker, C.; Ashley, C.; Harris, T. *Mat. Res. Soc. Symp. Proc.* **1998**, *519*, 183.
20. Brinker, C.; Scherer, G. *Sol-gel Science: The Physics and Chemistry of Sol-Gel Processing*; Academic Press: New York, NY, 1990.
21. Baek, J.; Kim, M.; Cha, H.; Lee, H.; Li, D.; Kim, J.; Kim, Y.; Moon, T.; Park, K. *Food Sci. Biotechnol.* **2003**, *12*, 639.

Indexes

Author Index

Brazzle, John D., 133
Brown, Suree S., 117
Bryant, Chet, 71
Ceremuga, Joseph T., 133
Cheng, Tu-Chen, 39, 233
Cole, Phillip J., 89
Coughlin, E. Bryan, 175
Cummings, Eric B., 133
Dai, Sheng, 117
Davalos, Rafael V., 133
Del Eckels, J., 39
Dong, Hua, 233
Fiechtner, Gregory J., 133
Fintschenko, Yolanda, 133
Geletii, Yurii V., 198
Hachman, John T., 133
Hadi, Masood Z., 39
Harmon, H. James, 57
Hart, Bradley R., 39
Hedden, Ronald C., 89
Hill, Craig L., 198
Hillesheim, Daniel A., 198
Hook, Gary, 71
Houser, Eric J., 71
Ilker, M. Firat, 175
Johnson-White, Brandy, 57
Kane, Staci R., 39
Kaner, Richard B., 101
Kanouff, Michael, 133
Lapizco-Encinas, Blanca H., 133
LaPuma, Peter, 71
Lawson, Glenn E., 3, 210
Lenhart, Joseph L., 89
Létant, Sonia E., 39

McGill, R. Andrew, 71
McGraw, Gregory J., 133
Mela, Petra, 133
Murray, G. M., 19
Okun, Nelya M., 198
Ong, Kate K., 233
Ott, Edward W., Jr., 19
Papantonakis, Mike R., 71
Perkins, Keith F., 71
Rastogi, Vipin K., 39
Reynolds, John G., 3, 39
Rondinone, Adam J., 117
Ross, Stuart K., 71
Shediac, Renee, 133
Shields, Sharon J., 39
Simmons, Blake A., 133
Simonson, Duane L., 71
Singh, Alok, 210
Snow, Eric S., 71
Southard, G. E., 19
Spangler, Brenda D., 159
Spangler, Charles W., 159
Stepnowski, Jennifer L., 71
Stepnowski, Stanley V., 71
Suo, Zhiyong, 159
Tarter, E. Scott, 159
Tew, Gregory N., 175
Unal, Burcu, 89
Van Houten, K. A., 19
Virji, Shabnam, 101
Wei, Yen, 233
Weiller, Bruce H., 101
Yin, Ray, 233
Yuan, Jian-Min, 233

Subject Index

A

Adhesion
  performance at low temperatures, 97–98
  tack, measurements, 91–92
  See also Non-aqueous polymer gels
Aerobic decontamination catalysts
  $Ag_2[(CH_3C(CH_2O)_3)_2V_6O_{13}]$, 204–206
  bis(2-chloroethyl)sulfide (HD) and VX decontamination, 199
  catalysis by nitrate/proton system, 203–204
  CEES (2-chloroethyl ethyl sulfide) mustard simulant to corresponding sulfoxide, 200–201
  CEES oxidation to CEESO by *t*-butyl hydroperoxide (TBHP), 205*f*
  experimental, 206–208
  nitrate/proton model system reactions, 207
  oxidative decontamination process for mustard (HD), 199–200
  sulfoxidation by $Fe^{III}[H(ONO_2)_2]PW_{11}O_{39}^{5-} \cdot HNO_3$, 201–202
  synthesis of $Ag_2[(CH_3C(CH_2O)_3)_2V_6O_{13}]$, 207
  UV-vis absorbance of oxidation of CEES to CEESO, 205*f*
  See also Chemical warfare agents
Agents, threats, 6*t*
Alpha particle
  CdSe/ZnS core/shell quantum dots (QDs), 123
  Ce-doped Y_2O_3 nanocrystals, 123–124
  detection by nanocrystalline inorganic scintillators, 122–125
  See also Scintillation detectors
Alpha particle detection, Ce-doped lanthanum phosphate nanocrystals, 124–125
Alteromonas species, hydrolytic catalytic activity of enzymes toward nerve agents, 52*t*
Amphiphilic polymers with antibacterial activity
  activity against bacterial vs. mammalian cells, 182–183
  advantages of ring-opening metathesis polymerization (ROMP)-based synthesis, 192
  antibacterial and hemolytic activities, 190–192
  applications, 188–189
  artificial liposomes as model membranes, 182, 183*f*
  catalysts for metathesis polymerization, 180*f*
  control experiments, 187–188
  decontamination, 13
  design based on norbornene derivatives, 176–177
  dye leakage experiment, 182, 183*f*
  effect of blood freshness, 192
  effect of membrane composition of liposomes, 186
  effect of polymer hydrophobicity, 186–187
  experimental considerations, 192
  homopolymerization studies, 180–181
  interactions with phopholipid membranes, 181–182
  lysis of anionic liposomes by various polymers, 183, 185

modular norbornene derivatives and resulting, 179t
monomer design based on norbornene derivatives, 176–177
monomer synthesis, 177–178
neutral zwitterionic liposomes preparation, 183, 184f
phospholipid membrane disruption activities, 181–185
polymer deprotection forming polyelectrolytes, 181
toxicity against mammalian cells, 182
Anthrax, detection needs, 160
Antibacterial activity
amphiphilic polymers with, 13, 188–189
cationic polymers, 176
See also Amphiphilic polymers with antibacterial activity
Antibodies
dendritic tethers for immobilization, 12
selective binding of analytes, 58
See also Dendritic tethers
Anti-terrorism, chemical and biological detection systems, 4
Aryl substituted silacyclobutanes, synthesis, 73–74
Attenuated total reflectance Fourier transform infrared (ATR/FTIR)
hydrogen bond acidic polycarbosilanes, 81–82, 86
nerve agent simulants, 82, 84f, 85f
sensor applications, 9

B

Bacterial cells
separation of different live, 134–135
spores, 136–137
vegetative cells, 136

See also Insulator-based dielectrophoretic devices
Beta particle
detection by nanoparticle liquid scintillator, 121–122
detection from ^{14}C, 122t
See also Scintillation detectors
Biological detection
dendritic tethers, 12
subject matter, 6
water-borne pathogens, 11–12
Biological threat agents, detection and decontamination, 5
Bioremediation program, U. S. Army, 234
Bioterrorism agents
detection needs, 160
See also Dendritic tethers
Bis(2-chloroethyl)sulfide
oxidative decontamination, 199–200
See also Aerobic decontamination catalysts
Bis(ethylhexyl)sebacate (BEHS)
swelling solvent, 91, 93f
See also Non-aqueous polymer gels
Bosch etch
fabrication of dielectrophoretic device, 137–138
See also Insulator-based dielectrophoretic devices

C

Carbon nanotube networks (CNN)
hydrogen bond acidic polycarbosilanes, 81, 83f
sensor applications, 9
Catalysts
aerobic decontamination of chemical warfare agents, 13–14
See also Aerobic decontamination catalysts; Chemical warfare agents

CdSe/ZnS core shell quantum dots (QDs), alpha particle detection, 123
Centers for Disease Control (CDC), partial list of threats, 6t
Cerium-doping
　lanthanum phosphate nanocrystals, 124–125
　Y_2O_3 nanocrystals, 123–124
Chemical, biological, nuclear and explosive (CBNE) materials, detection, 4
Chemical detection
　design approaches, 7
　enzyme-based photoluminescent porous silicon detector, 8–9
　features, 7
　molecularly imprinted polymer phosphonate sensors, 7–8
　nanoparticles in scintillation detectors, 11
　non-aqueous polymer gels with broad temperature performance, 10
　polyaniline nanofibers, 10–11
　sensors for reagentless detection of chemical analytes, 9
　sorbent hydrogen bond acidic polycarbosilanes, 9
　subject matter, 6
Chemical detectors, development of new materials, 6
Chemical sensor, hydrogen bond acidic polycarbosilanes, 9
Chemical warfare agents
　catalysts for aerobic decontamination, 13–14
　chemical structures, 5f
　decontamination, 199
　enzyme-based photoluminescent porous silicon detector, 8–9
　focus of book, 4–5
　list of threats, 6t
　See also Aerobic decontamination catalysts; Optical enzyme-based sensors; Photoluminescent (PL) porous silicon (PSi)
Chemical weapons
　bioremediation program, 234
　Geneva Protocol, 68
2-Chloroethyl ethyl sulfide (CEES)
　catalysis by $Ag_2[(CH_3C(CH_2O)_3)_2V_6O_{13}]$, 204–206
　catalysis by nitrate/proton system, 203–204
　catalysts for homogeneous air-based oxidation, 202t
　mustard simulant, 200
　oxidation to CEESO by t-butyl hydroperoxide (TBHP), 205f
　sulfoxidation, 200–201
　See also Aerobic decontamination catalysts
Cholinesterase surfaces
　detection limits for various analytes, 65t
　detection of organophosphonates, 64t
　enzymes acetylcholinesterase and butyrylcholinesterase, 65–67
　See also Optical enzyme-based sensors
Computational modeling
　dielectrophoretic device, 149–154
　See also Insulator-based dielectrophoretic devices
Coumaphos
　cholinesterase and/or organophosphorus hydrolase techniques, 64t
　detection limits, 65t
Crosslink density
　polybutadiene gels, 94
　See also Non-aqueous polymer gels
Cross-linking. See Phospholipid vesicles

D

Dansyl dye, emission spectra of functionalized porous silicon (PSi) surface, 45, 46f
Decontamination
  aerobic, of chemical warfare agents, 13–14
  amphiphilic polymers with potent antibacterial activity, 13
  nanocapsules from divinylbenzamide phosphoethanolamine, 14
  nanoencapsulation of organophosphorus acid anhydrolase (OPAA), 14–15
  oxidation, 13–14
  polynorbornene derivatives, 13
  subject matter, 6
  See also Aerobic decontamination catalysts
Dendritic tethers
  design and synthesis, 12
  design and synthesis of dithiol tethers, 160–165
  dithiol "flower" and "meadow" syntheses, 166–167, 168
  dithiol "flowers-in-the-meadow" mixed SAM, 167f
  dithiol immobilization agents with flexible design controls, 161f
  efficacy of dithiol "flower-in-the-meadow" self-assembled monolayer (SAM) design for warfarin binding, 167
  enhancing SAM stability, 165–167
  final assembly of immobilization reagent with tether and antibody functionality, 163–164
  hydrophobic or hydrophilic chains in antibody attachment segment, 161f
  introducing rigid rod segment to precursor, 162–163
  nonspecific binding characteristics, 167, 169f
  SAMs losing performance in biosensor applications, 165–166
  surface immobilized protein, 169f
  surface plasmon resonance (SPR) comparison, 169f
  synthesis of precursor, 162
  typical "flowers-in-the-meadow" SAM, 165, 166f
Design approaches, chemical detection, 7
Detection systems, chemical and biological, 4
DFP nerve agent
  hydrolytic catalytic activity of enzymes, 52t
  structure, 52
Diazinon
  cholinesterase and/or organophosphorus hydrolase techniques, 64t
  detection limits, 65t
Dibutyl phthalate (DBP)
  swelling solvent, 91, 93f
  See also Non-aqueous polymer gels
Dichlorvos, cholinesterase and/or organophosphorus hydrolase techniques, 64t
Dielectrophoretic devices
  water-borne pathogen monitoring and separation, 11–12
  See also Insulator-based dielectrophoretic devices
Diisopropyl fluorophosphate (DFP), cholinesterase and/or organophosphorus hydrolase techniques, 64t
Divinylbenzamide phosphoethanolamine nanocapsules from polymerizable, 14
  See also Phospholipid vesicles

Drug delivery, nanocapsules for encapsulation, 14

E

Encapsulation, decontamination, 14
Enzyme-based detectors
  catalytic rate, 58–59
  See also Optical enzyme-based sensors; Photoluminescent (PL) porous silicon (PSi)
Enzymes
  binding specificity, 58
  encapsulation into mesoporous molecular sieves, 235
  hydrolytic catalytic activity toward nerve agents, 52t
  nanoencapsulation with mesoporous materials, 14–15
  See also Nanoencapsulation of enzymes
Europium(III) ligand complexes
  background luminescence, 30–31, 32f
  binding site for organophosphates, 29
  characterization of β-diketone complexes, 27–28
  conversion to imprint complex, 27f
  β-diketone complexes, 25–27
  ligand monomer choice, 29–30
  luminescence titration method, 30
  polymer phosphonate sensors, 7–8
  reversible addition fragmentation transfer (RAFT) polymerization, 28f
  synthesis of tris(β-diketone) Eu(III) complex, 25f
Evanescent wave absorbance spectroscopy (EWAS)
  experimental setup, 61f
  measurement technique, 59
  schematic, 60f

See also Optical enzyme-based sensors

F

Flow cell
  development using functionalized porous silicon (PSi), 46–47
  sealed system for real-time fluorescence detection, 48f
Fluoroalcohols, enhancing polyaniline response to hydrazine, 109, 110f

G

G agents
  hydrolytic catalytic activity of enzymes toward nerve agents, 52t
  organophosphorus nerve agents, 40
  structures, 52
Geneva Protocol. chemical weapons, 68
Glass master, insulator-based dielectrophoretic devices, 12
Glucuronidase (GUC)
  activity of immobilized GUC, 48–49, 50f
  attachment to functionalized porous silicon (PSi) surface, 47–50
  fluorescence behavior of immobilized enzyme on PSi, 49–50, 51f
  scanning electron microscopy (SEM) of PSi surface, 49f
  See also Photoluminescent (PL) porous silicon (PSi)

H

Hemolytic activities, amphiphilic polymers, 190–192

Homeland defense, chemical and biological detection systems, 4
Homeland security
  polyaniline nanofibers, 10–11, 102–103
  See also Polyaniline nanofibers
Horseradish peroxidase (HRP), nanoencapsulation, 239
Hydrazine detection, conventional and nanofiber polyaniline films, 107, 109, 110f
Hydrofluoric acid (HF) etch
  fabrication of dielectrophoretic device, 137–138
  See also Insulator-based dielectrophoretic devices
Hydrogels
  classic poly-(N-isopropylacrylamide) (PNIPAM), 90
  See also Non-aqueous polymer gels
Hydrogen bond acidic polycarbosilanes
  attenuated total reflectance Fourier transform infrared (ATR–FTIR) vapor detection, 81–82, 86
  chemical sensors, 9, 72
  chemical structures of functionalized polymers (CS3PP2, NMA, and HC), 77f
  experimental, 73–75
  fluoroalcohol-substituted naphthalene model compounds preparation, 74–75
  Fourier transform infrared spectrum showing fluoroalcohol groups, 77, 78f
  FTIR difference spectra of diisopropyl methylphosphonate (DIMP) and dimethyl methylphosphonate (DMMP) sorbed by HC polymer, 85f
  FTIR spectra of DIMP and DMMP compared to HC polymer, 84f
  hexafluoroacetone (HFA) and polycarbosilanes reacting with alkenyl pendant groups, 74
  HFA and polycarbosilanes reacting with aryl pendant groups, 74
  linear, hyperbranched and dendritic structures, 75
  materials, 73
  parallel exposure of HC and NMA coated surface acoustic wave (SAW) devices, 80f
  pathway to fluoroalcohol substituted polycarbosilanes, 75–76
  properties of optimal material, 72
  representative monomers for parent polycarbosilanes, 76f
  SAW devices, 80–81
  schematic of TravelIR™ miniature ATR–FTIR spectrometer, 83f
  sensor measurements, 79–80
  silacyclobutane monomers, 75
  single wall carbon nanotube (SWNT) sensors, 81, 83f
  structures of primary substitution products from 2-ethylnaphthalene with HFA, 79f
  substituted silacyclobutane monomers to linear polycarbosilanes, 75
  substitution patterns of reaction of naphthalene-substituted polycarbosilanes with HFA, 77, 79
  synthesis of monomers, 73–74
  synthesis of parent polycarbosilanes, 74
  synthetic routes, 76f
  temporal response of SWNT chemicapacitors, 83f
  thermal stability, 77
  threat signature vapors, 72

Hydrogen sulfide
  detection by polyaniline films, 111
  metal sulfides enhancing detection by polyaniline films, 112, 113*f*
Hydrosilation, polycarbosilanes preparation, 75
Hydrosilation route, Lewis acid
  FTIR and fluorescence spectra of porous silicon (PSi) before and after, 45*f*
  linker system for porous silicon (PSi) surface, 42–43, 44*f*

I

Immobilization
  antibodies for pathogen detection, 12
  enzymes for optical biosensors, 61–63
  See also Dendritic tethers
Inorganic nanocrystals. *See* Scintillation detectors
Insulator-based dielectrophoretic devices
  application and theory, 134
  channel with 4 x 10 post configuration, 146*f*
  computational modeling comparison of device performance, 149–154
  determination of trapping thresholds, 139
  device fabrication by hydrofluoric acid (HF) etch and reactive ion (Bosch) etch, 137–138, 154–155
  device performance characterization, 145–146
  differential trapping comparison of biological tracers, 145, 148*f*
  dimensionless electric field, 140, 149–151
  electric potential vs. distance along longitudinal centerlines of Bosch and HF channels, 149, 150*f*
  experimental, 135–141
  finite element meshes of HF and Bosch iDEP channels, 149*f*
  inert particles, 136
  interferometer data of stamp and replicates for posts for Bosch and wet HF replication, 144*f*
  metrology, 141–143
  metrology system, 135–136
  modeling protocols, 139–141
  particle trajectories, 140, 151–152, 153*f*
  particle velocities, 140–141, 152, 153*f*, 154*f*
  photographs of polystyrene (PS) beads in array from Bosch etch-derived device, 147*f*
  polymer microfluid devices, 135
  replication process measurements by Bosch etch and HF glass etch, 143*t*
  safety considerations, 139
  schematic of experimental setup, 146*f*
  schematic of metrology measurements, 142*f*
  separating species of live bacterial cells, 134–135
  separation demonstration of PS beads by size, 147*f*
  spores, 136–137
  threshold trapping, 139
  threshold trapping electric potential differences for Bosch and HF channels, 154, 155*f*
  tracers and background solutions, 136–137
  trapping thresholds by Bosch and HF etch device types, 145–146, 148*f*
  vegetative cells, 136

J

Joint Chemical Agent Detector (JCAD), surface acoustic wave (SAW), 68

L

Lanthanide ion Eu^{3+}. *See* Molecularly imprinted polymers (MIPs)
Lanthanum phosphate nanocrystals, alpha particle detection by Ce-doped, 124–125
Lewis acid hydrosilation route
  FTIR and fluorescence spectra of porous silicon (PSi) before and after, 45f
  linker system for porous silicon (PSi) surface, 42–43, 44f
Liposomes
  disruption of neutral and anionic, 182–183
  effect of membrane composition of, 186
  illustration of, and dye leakage experiment, 182, 183f
  lysis of anionic, by amphiphilic polymers, 185f
  preparation of neutral zwitterionic, as mimics for mammalian cell membranes, 183, 184f
  See also Amphiphilic polymers with antibacterial activity
Lithium-6 phosphate nanocrystals, slow neutron detection, 125–126
Living ring opening metathesis polymerization (ROMP), amphiphilic polymers, 13
Luminescence detection
  synthesis of imprinted polymers, 20–21
  titration method for europium(III) complexes, 30
  See also Molecularly imprinted polymers (MIPs)

M

Malathion
  cholinesterase and/or organophosphorus hydrolase techniques, 64t
  detection limits, 65t
Maleic anhydride (MA), functional groups on polybutadiene, 91, 93f
Mesoporous materials
  nanoencapsulation of organophosphorus acid anhydrolase (OPAA), 14–15
  pore volume, surface area, and controlled pore size, 239–240
  preparation, 236–237
  properties of OPAA in, 238t
  See also Nanoencapsulation of enzymes
Metal sulfides, enhancing hydrogen sulfide detection by polyaniline films, 112, 113f
Methamidophos, cholinesterase technique, 64t
Microfabrication
  devices using hydrofluoric acid (HF) etch and reactive ion etch (Bosch), 137–138
  insulator-based dielectrophoretic devices, 11–12
  See also Insulator-based dielectrophoretic devices
Modeling
  dielectrophoretic device, 149–154
  See also Insulator-based dielectrophoretic devices
Molecularly imprinted polymers (MIPs)
  background luminescence, 30–31, 32f

calibration curve for reversible addition fragmentation transfer (RAFT) pinacolyl methylphosphonate (PMP) MIP, 34f
characterization of β-diketone complexes, 27–28
chemical detection, 7–8
conventional preparation, 19
conversion to imprint complex, 27f
detection mechanism for PMP, 20–21
β-diketone complexes of lanthanides, 25–27
dithiobenzoic acid 1-(4-(4,4,4-trifluorobutane-1,3-dione)-naphthalen-1-yl)-ethyl ester (HDBNTFA) synthesis, 22–23
Eu^{3+} ion luminescence detection, 20–21
experimental, 22–28
general procedure for RAFT polymerizations with methacrylics, 24
general procedure for RAFT polymerizations with styrenics, 24–25
improving Eu(III) binding site for organophosphates, 29
instrumentation, 22
ligand monomer choice, 29–30
luminescence spectra of $Eu(NTFA)_3$ with addition of PMP in chloroform, 31f
luminescence spectra of tris(divinyldibenzoylmethanato) europium(III), 29f
luminescence titration method, 30
luminescence vs. equivalents of PMP added, 32fr
mixed ligand approach, 31, 33
polymer synthesis, 24–25
preparation of dithioester ligand, 24f
preparation of tris chelate, 26f
RAFT polymerization of, for PMP, 28f
RAFT polymers, 33, 36
scanning electron microscopy (SEM) of large particle of mixed ligand polymer, 35f
SEM of large particle of RAFT polymer using methoxyethanol, 36f
SEM of large particle of RAFT polymer using toluene, 35f
synthesis for luminescence detection, 20–21
synthesis of $DBNTFA_3Eu·3H_2O$, 23
synthesis of tris(β-diketone) europium(III) complex, 25f
synthesis of vinyl-β-diketones, 23f
Mustard
oxidative decontamination, 199–200
See also Aerobic decontamination catalysts

N

Nanocapsules
polymerizable divinylbenzamide phosphoethanolamine, 14
See also Phospholipid vesicles
Nanocrystals, inorganic
alpha particle detection, 122–125
slow neutron detection, 125–126
See also Scintillation detectors
Nanoencapsulation of enzymes
activities of nanoencapsulated organophosphorus acid anhydrolase (OPAA) and reusability, 242t
Alteromonas sp. JD6.5 OPAA, 239
aqueous solution, 241, 243
comparing nanoencapsulated OPAA activity in different solvents, 244f

deleterious effect of methanol on free enzyme, 246, 248f
denaturing effect of free OPAA, 244, 246
experimental, 236–239
horseradish peroxidase (HRP), 239
low volume shrinkage process, 240–241
materials, 236
mesoporous material preparation, 236–237
mesoporous molecular sieves, 235
mixed aqueous-organic solution, 243–244, 246
nanoencapsulated-enzyme polymers, 237, 238t
nanoencapsulated OPAA activity in various solvents, 245t
OPAA activity determination, 237, 239
organophosphorus acid anhydrolase (OPAA), 14–15
percent enzymatic activity to free OPAA vs. gelation time, 240f
pore volume, surface area and controlled pore size, 239–240
protective advantage for enzyme in dimethyl formamide (DMF), 246, 247f
regeneration wash procedure, 243
reusability, 247–248, 250
reusability activity summary of free and entrapped OPAA, 249t
wash and transfer process, 247
Nanofibers
  polyaniline, for toxic chemical detection, 10–11, 102–103
  See also Polyaniline nanofibers
Nanoparticles
  scintillation detectors, 11
  See also Scintillation detectors
National security, chemical, biological, nuclear and explosive (CBNE) materials, 4

Nerve agents
  chemical, 5
  hydrolytic catalytic activity of enzymes, 52t
Neutron detection
  lithium-6 phosphate nanocrystals, 125–126
  procedure, 120
  See also Scintillation detectors
p-Nitro-phenyl-soman, chemical warfare agent, 41
Non-aqueous polymer gels
  adhesive performance at low temperatures, 97–98
  bis(ethylhexyl)sebacate (BEHS) and dibutyl phthalate (DBP) solvents, 91, 93f
  broad temperature performance, 10
  constituents of gel formation, 92, 93f
  cross-link density, 94
  DBP vs. BEHS as solvent candidate, 94–95
  dependence of mass uptake of solvent on solvent solubility parameter, 92, 93f
  experimental, 91–92
  loss modulus vs. temperature for, with varying maleic anhydride content, 94, 95f
  material requirements for broad temperature performance, 90–91
  polybutadiene monomers, 91, 93f
  rheological measurement method, 91
  room temperature swelling, 96
  small angle neutron scattering (SANS), 91, 96–97
  solvent quality and polymer-solvent phase behavior, 97
  storage modulus vs. temperature for, with varying maleic anhydride content, 94, 95f

storage modulus vs. temperature for MA5 gels with varying amount of BEHS and DBP, 96f
swelling of crosslinked polybutadiene in solvents, 92, 93f
tack adhesion measurements, 91–92

Norbornene derivatives
catalysts for metathesis polymerization, 180f
examples and resulting amphiphilic polymers, 179t
general structures of amphiphilic modular, 177f
monomer synthesis, 177–178
preparation schematic, 178f
starting point for monomer design, 176–177
See also Amphiphilic polymers with antibacterial activity

O

Optical enzyme-based sensors
acetylocholinesterase terminating nerve impulse transmission, 65–66
analysis of chemical weapons, 68
cholinesterase surfaces, 65–67
detected organophosphonates using cholinesterase and/or organophosphorus hydrolase based techniques, 64t
detection limits for various analytes, 65t
evanescent wave absorbance spectroscopy (EWAS), 59–61
experimental setup for EWAS, 61f
factors affecting rate of catalysis, 58–59
immobilization process, 61–63
interaction of analyte with porphyrin-enzyme complex, 60f
methods, 59–63
organophosphorus hydrolase surfaces, 63–65
reagentless detection of chemical analytes, 9
real-time detection technique, 67
reversible inhibition of enzyme by association-sensitive colorimetric agent, 67
specificity, 58
standard absorption measurement, 60f
systems based on reversible, competitive inhibition of cholinesterases by porphyrin, 66

Organophosphonates (OPs), anti-cholinesterase activity, 58

Organophosphorus acid anhydrolase (OPAA)
activity determination, 237, 239
Alteromonas sp. strain JD6.5, 234
attachment of *Alteromonas* sp. JD6.5 OPAA-2 to photoluminescent porous silicon, 50, 52–54
commercialization and marketing for OPAA, 234–235
emission of OPAA-functionalized PSi before exposure to p-nitro-phenyl-soman, 53, 54f
etching methods for modifying porous silicon (PSi) surfaces for, 42f
methanol effect on free enzyme, 246, 248f
nanoencapsulated-enzyme polymers, 238t
nanoencapsulation, 14–15
p-nitro-phenol production vs. time by OPAA-functionalized PSi device, 52, 53f
protective advantage of entrapped OPAA in dimethyl formamide (DMF), 246, 247f

See also Nanoencapsulation of enzymes
Organophosphorus hydrolase surfaces
detection of organophosphonates, 64t
See also Optical enzyme-based sensors
Organophosphorus nerve agents, G agents, 40
Organophosphorus pesticides, chemical warfare agents, 5
Oxidation catalysts
decontamination of chemical warfare agents, 13–14
See also Aerobic decontamination catalysts

P

Paraoxon
cholinesterase and/or organophosphorus hydrolase techniques, 64t
detection limits, 65t
Parathion, cholinesterase and/or organophosphorus hydrolase techniques, 64t
Pathogen detection, immobilization of antibodies, 12
Pathogen separation. See Insulator-based dielectrophoretic devices
Phosphoethanolamine, nanocapsules from polymerizable divinylbenzamide, 14
Phospholipid membranes
artificial liposomes as model membranes, 182
dye leakage experiment, 182, 183f
interactions of amphiphilic macromolecules, 181–182
See also Amphiphilic polymers with antibacterial activity

Phospholipid vesicles
carboxyfluorescein entrapment and linkage from vesicles, 217
characterization of phospholipid physical properties, 215
cross-linked vesicles, 221–222
cross-linking and properties, 212
cross-linking experiments, 216
diameter of cross-linked vesicles after solvent treatment, 227t
differential scanning calorimetry (DSC), 222, 224t
efficiency and extent of cross-linking, 218–219, 220f
entrapment of oil red-O in freeze-dried cross-linked vesicles, 227
experimental, 212–218
graphic representation of vesicle cross-linking, 220f
^{1}H NMR of, as function of ultraviolet irradiation time, 223f
hydrophobic dye encapsulation in cross-linked vesicles, 217–218
leakage of carboxyfluorescein from, 227, 230f
preparation and characterization of vesicles, 215–218
schematic of cross-linked vesicle, 213f
stability of cross-linked vesicles, 216–217, 225–226
surface pressure-area isotherms, 219, 220f
synthesis of phospholipids and intermediates, 213–214
TEM (transmission electron microscopy) of vesicles, 228f, 229f
TEM visualization, 225
thermotropic behavior of hydrated phospholipids, 222, 224–225
transition properties, 224t
vesicle dispersion, 220–221
visualization of vesicles, 215–216

See also Nanocapsules
Photoluminescent (PL) porous silicon (PSi)
  activity of immobilized glucuronidase (GUC), 48, 50*f*
  attachment of *Alteromonas* sp. JD6.5 OPAA–2, 50, 52–54
  attachment of GUC, 47–50
  characterization of functionalized PSi surface, 43, 45–46
  detectors for chemical warfare agents, 8–9
  development of flow cell, 46–47
  emission of OPAA-functionalized PSi before exposure to p-nitro-phenyl-soman, 53, 54*f*
  emission spectra of PSi functionalized with dansyl dye, 46*f*
  etching methods modifying PSi surfaces, 42*f*
  experimental, 41
  fluorescence behavior of immobilized enzyme on PSi, 49–50, 51*f*
  Fourier transform infrared and fluorescence spectra of PSi before and after Lewis acid hydrosilation, 45*f*
  GUC fastened to functionalized PSi surface, 49*f*
  hydrolytic catalytic activity of enzymes toward nerve agents, 52*t*
  immobilization linker system, 44*f*
  linker system by Lewis acid hydrosilation route, 42–43, 44*f*
  Michaelis–Menten equation for activity, 48–49, 50*f*
  p-nitro-phenol production vs. time by OPAA-functionalized PSi device, 52, 53*f*
  p-nitro-phenyl soman, 41
  sealed flow cell for real-time fluorescence, 48*f*
  trypsin digest of PSi surface with biotin-streptavidin, 46, 47*f*
  varying porosity and optical properties, 41–42
Pinacolyl methylphosphonate (PMP)
  mechanism for detection, 20–21
  See also Molecularly imprinted polymers (MIPs)
p-nitro-phenyl-soman, chemical warfare agent, 41
PNP (p-nitro-phenyl), production vs. time by enzyme-functionalized porous silicon device, 52, 53*f*
Polyaniline
  comparing response of, and polyaniline nanofibers to HCl, 106, 107*f*
  conventional, 102
  doped, 105–106
  film thickness, 106
  fluoroalcohols enhancing response to hydrazine, 109, 110*f*
  hydrazine detection, 107, 109, 110*f*
  response time of films, 107, 108*f*
  reversible acid/base doping/dedoping process, 105
Polyaniline nanofibers
  comparing conventional polyaniline and, films to HCl, 106, 107*f*
  composites with polyaniline, 111
  detecting hydrogen sulfide, 111
  experimental, 103, 105
  film thickness, 106
  fluoroalcohols enhancing response to hydrazine, 109, 110*f*
  future work, 113
  hydrazine detection, 107, 109, 110*f*
  measuring acid concentration, 106
  outperforming conventional polyaniline films, 112
  resistance changes of, films with metal sulfides on exposure to H_2S, 112, 113*f*
  response time, 107, 108*f*

reversibility of, films exposed to
bases, 106–107, 108f
reversible acid/base
doping/dedoping process,
105
scanning electron microscope
(SEM) image, 104f
solubility product constants for
metal sulfides, 112t
toxic chemical detection, 10–11
transition electron microscope
(TEM) image, 104f
Polybutadiene gels. See Non-aqueous
polymer gels
Polycarbosilanes
chemical sensor applications, 9
monomers for preparation of
parent, 76f
preparation of fluoroalcohol
substituted, 75–76
reaction of naphthalene-substituted,
with hexafluoroacetone (HFA),
77, 80
synthetic routes, 76f
See also Hydrogen bond acidic
polycarbosilanes
Polymer gels
applications, 90
classic hydrogel of poly-(N-
isopropylacrylamide)
(PNIPAM), 90
temperature performance for non-
aqueous, 10
See also Non-aqueous polymer gels
Poly(N-isopropylacrylamide)
(PNIPAM)
classic hydrogel system, 90
See also Non-aqueous polymer gels
Polynorbornene derivatives,
antibacterial activity, 13
Polyoxometalates (POMs)
decontamination catalysts, 200
See also Aerobic decontamination
catalysts

Porous silicon (PSi). See
Photoluminescent (PL) porous
silicon (PSi)
Porphyrins
colorimetric indicators in detection
protocols, 63
interaction of analyte with
porphyrin-enzyme complex, 60f
optical detection, 59
See also Optical enzyme-based
sensors

Q

Quantum dots (QDs)
CdSe/ZnS core/shell QDs, 123
inorganic nanocrystals, 118
See also Scintillation detectors
Quartz crystal microbalance (QCM)
tether attachment to surface, 161
See also Dendritic tethers

R

Radiation
detection methods, 120
methods for detecting ionizing, 118
neutron measurements, 120
See also Scintillation detectors
Reactive ion (Bosch) etch
fabrication of dielectrophoretic
device, 137–138
See also Insulator-based
dielectrophoretic devices
Reagentless detection, optical
enzyme-based sensors, 9
Reusability
activities of nanoencapsulated
organophosphorus acid
anhydrolase (OPAA), 241, 242t
activities of reused
nanoencapsulated OPAA in

various solvent solutions, 247–248, 250
free and entrapped OPAA in aqueous-organic solvent, 249*t*
regeneration wash procedure, 243
wash and transfer process, 247
Reversible addition fragmentation transfer (RAFT) polymerization
pinacolyl methylphosphonate (PMP) molecularly imprinted polymer (MIP), 33, 34*f*, 36
polymer phosphonate sensors, 7–8
See also Molecularly imprinted polymers (MIPs)
Rheology
measurements, 91
polybutadiene gels, 94–96
Ring opening metathesis polymerization (ROMP)
advantages of ROMP-based synthesis, 192
amphiphilic polymers, 13
homopolymerization studies, 180–181
metathesis catalysts, 180*f*
monomers based on norbornene derivatives, 176–177
monomer synthesis, 177–178
See also Amphiphilic polymers with antibacterial activity

S

Sarin (GB agent)
detection limits, 65*t*
organophosphate development and use, 58
structure, 52
Scintillation detectors
alpha particle detection by nanocrystalline inorganic scintillators, 122–125

beta particle detection by 2,5-diphenyloxazole (PPO), 1,4-bis-2-(5-phenyloxazolyl)-benzene (POPOP)-doped polystyrene nanoparticle liquid scintillator, 121–122
CdSe/ZnS core/shell quantum dots (QDs), 123
Ce-doped lanthanum phosphate nanocrystals, 124–125
Ce-doped Y_2O_3 nanocrystals, 123–124
detection of beta particles from ^{14}C, 122*t*
experimental, 119–120
ideal material, 118
nanoparticles, 11
neutron measurements procedure, 120
pulse height spectra from alpha particle detection by $La_{0.9}Ce_{0.1}PO_4$ nanocrystals, 125*f*
pulse height spectra from neutron detection by Li-6 phosphate nanocrystals, 127*f*
radiation detection methods, 120
slow neutron detection with Li-6 phosphate nanocrystals as neutron absorbers, 125–126
transmission electron micrograph of Li-6 phosphate nanocrystals, 126*f*
Self-assembled monolayers (SAMs)
enhancing stability, 165–167
typical "flower-in-the-meadow", 165, 166*f*
See also Dendritic tethers
Semiconductor nanocrystals
inorganic nanocrystals, 118
See also Scintillation detectors
Sensors
design and specificity, 58
hydrogen bond acidic polycarbosilanes, 9, 79–80

optical enzyme-based, for chemical analytes, 9
polyaniline nanofibers, 103
Signature vapors, identifying threats, 72
Silacyclobutanes
   aryl substituted, preparation, 73–74
   See also Hydrogen bond acidic polycarbosilanes
Silicon detector
   enzyme-based photoluminescent, 8–9
   See also Photoluminescent (PL) porous silicon (PSi)
Silicon master, insulator–based dielectrophoretic devices, 12
Simulants, chemical warfare agents, 5
Single wall carbon nanotube network (SWNT-CNN) sensors, hydrogen bond acidic polycarbosilanes, 81, 83f
Small angle neutron scattering (SANS), polymer gels, 91, 96–97
Sol-gel process, nanoencapsulation of organophosphorus acid anhydrolase (OPAA), 14–15
Solvent swelling, crosslinked polymers, 90
Soman (GD agent), structure, 52
Special nuclear material (SNM) detection
   fundamental research, 118
   nanoparticles, 11
Specificity, sensor design, 58
Sulfur mustard, decontamination by oxidation, 13–14
Surface acoustic wave (SAW)
   development of new materials, 6
   hydrogen bond acidic polycarbosilanes, 80–81
   Joint Chemical Agent Detector (JCAD), 68
   sensor applications, 9
Surface plasmon resonance (SPR)
   efficacy of self-assembled monolayer design using SPR biosensor, 167
   tether attachment to surface, 161
   See also Dendritic tethers
Swelling network, polybutadiene gels, 92, 93f

T

Tabun (GA agent), structure, 52
Tack adhesion
   measurements, 91–92
   polymer gel at low temperatures, 97–98
   See also Non-aqueous polymer gels
Temperature performance
   non-aqueous polymer gels, 10
   See also Non-aqueous polymer gels
Thermal stability, hydrogen bond acidic polycarbosilanes, 77
Toxins, threats, 6t
Trichlorfon, cholinesterase technique, 64t

U

United States Army, bioremediation program, 234

V

Vapors, identifying threat signature, 72
Vesicles
   applications, 211
   nanocapsules for encapsulation of materials, 14
   See also Phospholipid vesicles
VX nerve agent
   decontamination by oxidation, 13–14

hydrolytic catalytic activity of
enzymes, 52*t*
oxidative decontamination, 199
structure, 52
See also Aerobic decontamination
catalysts

W

Warfarin, efficacy of self-assembled
monolayer design using surface
plasmon resonance biosensor, 167
Water-borne pathogens

insulator-based dielectrophoretic
devices, 11–12
See also Insulator-based
dielectrophoretic devices
Weapons of mass destruction (WMD),
chemical, biological, nuclear and
explosive (CBNE) materials, 4

Y

Y_2O_3 nanocrystals, alpha particle
detection by Ce-doped, 123–
124